T0304740

FLORA & FAUNA HANDBOOK NO. 3

THE POTATO BEETLES

The Genus Leptinotarsa in North America
(Coleoptera: Chrysomelidae)

FLORA & FAUNA HANDBOOKS

This series of handbooks provides for the publication of book length working tools useful to systematics for the identification of specimens, as a source of ecological and life history information, and for information about the classification of plant and animal taxa. Each book is separately numbered, starting with Handbook No. 1, as a continuing series. The books are available by subscription or singly.

Each book treats a single biological group of organisms (e.g., family, subfamily, single genus, etc.) or the ecology of certain organisms or certain regions. Catalogs and checklists of groups not covered in other series are included in this series.

The books are complete by themselves, not a continuation or supplement to an existing work, or requiring another work in order to use this one. The books are comprehensive, and therefore, of general interest.

Books in this series to date are:

Handbook No. 1—THE SEDGE MOTHS, by John B. Heppner

Handbook No. 2—INSECTS AND PLANTS: Parallel Evolution and Adaptations, by Pierre Jolivet

Handbook No. 3—THE POTATO BEETLES, by Richard L. Jacques, Jr.

Handbook No. 4—THE PREDACEOUS MIDGES OF THE WORLD, by Willis W. Wirth and William L. Grogan, Jr.

THE POTATO BEETLES

The Genus Leptinotarsa in North America

(Coleoptera: Chrysomelidae)

By

RICHARD L. JACQUES, JR.

Professor of Biology
Fairleigh Dickinson University
Rutherford, New Jersey

LONDON AND NEW YORK

Acquisitions and Production Editor: Ross H. Arnett, Jr.

First published 1988 by E. J. Brill

2 Park Square, Milton Park, Abingdon, Oxfordshire OX14 4RN
52 Vanderbilt Avenue, New York, NY 10017

Routledge is an imprint of the Taylor & Francis Group, an informa business

First issued in hardback 2019

A FLORA & FAUNA HANDBOOK

Library of Congress Cataloging in Publication Data:

Jacques, Richard L. 1945-1988
The potato beetles.

(Flora & Fauna handbook; no. 3)
Bibliography: p.
Includes indexes.
1. Leptinotarsa—North America. 2. Insects—North America. I. Title. II. Series.
QL596.C5J33 1987 595.76'4 87-15095
ISBN 0-916846-407

ISBN 13: 978-0-916846-40-4 (pbk)
ISBN 13: 978-1-138-42374-9 (hbk)

PREFACE

The study of the North American (including Mexico) *Leptinotarsa* (Coleoptera: Chrysomelidae), presented herein is the culmination of a number of years of study of the taxonomy and biology of the 31 species of Potato Beetles in this area, members of the genus *Leptinotarsa*. This genus is strictly a New World genus except for the one pest species that has been transported on potatoes throughout the world. Research for the revision was conducted primarily as part of the requirements for the Doctor of Philosophy degree at Purdue University, Department of Entomology; the degrees was awarded in 1972.

This work includes studies on 31 of the 41 known species of *Leptinotarsa* of the world. Included are host records, when available, and distribution data. The most complete information is available for the nine species found in the United States. Considerable work remains to be done on the species of *Leptinotarsa*, including the finding of additional information on host plants, larval descriptions, predators, and life cycles. It would be ideal if we could have as much information about all of the species as we know about the Colorado Potato Beetle, *L. decemlineata*, and we are still finding new information about this, the most famous of all the *Leptinotarsa*ı

The name "Potato Beetles" is used to serve both the public interested in these species, and professional entomologists as a "benchmark" for the group. Most individuals are familiar with the Colorado Potato Beetle and it is hoped that "Potato Beetles" used as a common name for the genus will heighten awareness of the presence of *Leptinotarsa* species in the United States.

This study of the North American species of *Leptinotarsa* is presented to allow relative ease and certainty, the identification of the North American species for all interested persons. Further, it will permit those persons to seek out and record additional data on this group.

Acknowledgements

Research for this work was initially conducted at Purdue University, West Lafayette, Indiana. The Department of Entomology provided support and facilities for this study. The National Science Foundation provided a dissertation enhancement grant in 1971 for field research in Arizona. The work was done under the direction of Ross H. Arnett, Jr., Ph.D., as major advisor. Many thanks must be extended to Ross Arnett for his persistence in making sure the taxonomic revision of *Leptinotarsa* was finally updated and published. His continued guidance and support for this and other projects is greatly appreciated.

Special mention must be extended to Gary Bernon, Ph.D., U.S. Department of Agriculture, APHIS, Otis Methods Development Laboratory, Otis, Massachusetts. It was Gary's invitation to a Colorado Potato Beetle workshop at the New Jersey Department of Agriculture in Trenton that renewed my interest in *Leptinotarsa*. His continued sharing of vital information on United Stated *Leptinotarsa* is greatly appreciated.

Thanks is also extended to Patrick Logan, Department of Plant Pathology and Entomology, University of Rhode Island, Kingston, Rhode Island, for providing valuable locality, host, and parasite/predator data.

Special thanks go to Arwin Provonsha, Curator of the Entomology Collection at Purdue University for his drawing of beetle elytra and genitalia, and to Adelaide Murphy of New York City for her illustrations of host plants and the cover and backpiece.

Study specimens were obtained on loan from the following institutions with the abbreviations used for collections (Arnett and Samuelson, 1986).

AMNH	American Museum of Natural History
CASC	California Academy of Science Collection
CISC	University of California at Berkeley Collection
FSCA	Florida State Collection of Arthropods
HAHC	Henry and Ann Howden Collection
INHS	Illinois Natural History Collection
ISUC	Iowa State University Collection
LACM	Los Angeles County Museum of Natural History
OSUC	Ohio State University Collection
PURC	Purdue University Research Collection
RLJC	Richard L. Jacques Collection (now at FSCA)
TAMC	Texas A & M University Collection
UCDC	University of California at Davis Collection
USNM	United States National Museum
USUC	Utah State University Collection

Finally, I would like to thank G. Lansing Blackshaw, Dean of the College of Science & Engineering at Fairleigh Dickinson University for his support in providing funds for additional illustrations and some publication costs.

Richard L. Jacques, Jr.
Rutherford, New Jersey
March 1987

CONTENTS

ABSTRACT

This study was initiated in 1970. Its objective is a taxonomic revision of the genus *Leptinotarsa* in North America.

The work is based on over 2,500 specimens representing all the species of *Leptinotarsa* in Canada, the United States, and Mexico. External morphological characters and the male genitalia were studied.

In Canada, the United States, and Mexico the genus *Leptinotarsa* is now represented in 31 species. Only one species is recorded from Canada, 12 species are recorded from the United States, and 29 species are recorded from Mexico. Three species are found exclusively in the United States. The genus *Leptinotarsa* is restricted to the Western Hemisphere except for *Leptinotarsa decemlineata* (Say) which has migrated to Europe and Asia.

Twelve species are placed in synonymy. A key to the species of *Leptinotarsa* of North America is presented. Biological and distributional data are presented with each species redescription. Total length, greatest width, interocular distance, head width, pronotum length and width, elytra length and width are recorded for each species. Range and standard deviations for measurements are included.

CHAPTER 1

THE CLASSIFICATION OF THE POTATO BEETLES

The family Chrysomelidae, or leaf beetles, is the third largest family of Coleoptera with 35,000 recorded species worldwide and estimates of an additional 25,000 unrecorded species yet to be discovered! As the common name suggests, these beetles feed on plants both in the larval and adult stages. Chrysomelids are known to feed on all portions of plants but especially the leaves. A majority of the highly evolved species, such as the *Leptinotarsa*, feed on leaves both as larva and adults. Both life stages feed on the same or related food plant.

The family Chrysomelidae can be identified on the basis of the following characteristics: body elongated, cylindrical to oval and flat; many species brightly colored. Tarsomeres 5-5-5, third tarsomere enlarged, fourth very small, sometimes minute, surrounded by dorsal depression of the third; antennae short, usually filiform, sometimes serrate or with a slight club (clavate), their base not, or on rare occasions slightly, surrounded by eyes; legs usually short; hind femora of many species (i.e. flea beetles) enlarged and fitted for jumping.

In the United States north of Mexico, Arnett (1985) records 1481 species of chrysomelids in 188 genera. The Chrysomelinae is one of the larger subfamilies in the order Coleoptera and the family Chrysomelidae. Worldwide there are some 2,000 species; in the United States north of Mexico there are 125 species in 15 genera in 7 tribes. (Table I)

The subfamily Chrysomelinae can be identified by the following characteristics; round or oval, convex; usually brightly colored. Head inserted into the prothorax to the eyes, eyes feebly emarginate; antennae moderately long, apical segments somewhat enlarged; antennal insertions widely separated. Prothorax usually broad and convex; side margins well defined, frequently emarginate in front. Procoxae transverse, widely separated; third segment of the tarsi entire, not bilobed. Elytra convex, covering abdomen; epipleural fold well defined.

The genus *Leptinotarsa* is assigned to the tribe Doryphorini, (= Zygogrammini) a tribe represented by 5 genera and 75 species in the United

TABLE I. Tribes in the subfamily Chrysomelinae with number of genera and species in the United States north of Mexico (Arnett 1985).

TRIBE	NUMBER OF GENERA	NUMBER OF SPECIES
Timarchini	1	3
Doryphorini	5	75
Chrysomelini	3	29
Phratorini	3	29
Entomoscelini	2	2
Prasocurini	2	5
Gonioctenini	1	4
7 Tribes	15	125

States. (Table II). Three of the genera have been recently revised. The genus *Labidomera* Chevrolat with 2 species in the United States was treated by Brown (1961). The genus *Calligrapha* Chevrolat includes 35 species and was discussed by Brown (1945) and Monros (1955). Brown (1962) also reviewed the genus *Chrysolina* Motschulsky which includes 16 species in the United States. *Zygogramma* Chevrolat and *Leptinotarsa* Stal was last considered by Linell (1896). This was followed by extensive evolutionary studies of *Leptinotarsa* by Tower (1906, 1918). Wilcox (1972) published a review of all the North American Chrysomeline leaf beetles which included a key to the *Leptinotarsa* of the United States.

Species of Doryphorini can be identified on the basis of the following characteristics; round to oval, convex; usually brightly colored. Procoxal cavities open behind; claws simple, separate at base, usually divergent; apical segment of maxillary palpus various.

TABLE II. Genera in the tribe Doryphorini with number of species in the United States north of Mexico (Arnett 1985).

GENUS	NUMBER OF SPECIES
Zygogramma Chevrolat	13
Leptinotarsa Stal	9
Labidomera Chevrolat	2
Calligrapha Chevrolat	35
Chrysolina Motschulsky	16
5 Genera	75

Most species of *Leptinotarsa* are not well known except for two United States species. *Leptinotarsa decemlineata* (Say), the Colorado potato beetle, is the best known member of this genus. It is well known both in the United States, where it was first described, and in Europe and Asia. It was introduced into Europe in 1875 and has since spread to Asia. The economic loss to potato crops worldwide has been severe and many entomologists would rate the Colorado potato beetle as one of the most important economic insects affecting human populations.

Another species of *Leptinotarsa* that is known in the southeastern United States is *Leptinotarsa juncta* Germar, once known as the False Colorado potato beetle and more appropriately known as the Horse-nettle beetle, named after its host plant *Solanum carolinense*. Other than these two species, the other 7 species in the United States as well as the other species in the genus are not well known. All the species of the genus except for the Colorado potato beetle are limited to the Western Hemisphere. Species can be found as far north as southern Canada and as far south as Brazil and Peru. In the United States only 2 of the 9 species are located outside the southwest United States. Most United States *Leptinotarsa* occur in an area from southern Texas around Brownsville west to New Mexico and Arizona. There are a total of 41 known species of *Leptinotarsa*, (Table III), 31 occur in Mexico and the United States, 9 occur in Central and South America and as stated before 9 of the 41 species occur in the United States.

TABLE III. Species of *Leptinotarsa*. List of species of *Leptinotarsa* of North America, including Mexico, Central America, and South America.

List of species of *Leptinotarsa* of North America, including Mexico, Central America, and South America

NAME	YEAR DESCRIBED	DISTRIBUTION
Leptinotarsa Star, 1858		
Polygramma Chevrolat, 1836		
Myocoryna Stal, 1858		
behrensi Harold	1877	Mexico, U.S.A.
modesta Jacoby	1883	
puncticollis Jacoby	1883	
belti Jacoby	1879	Guatemala, Nicaragua
boucardi Archard	1923	Mexico
cacica Stal	1858	Mexico

calceata Stal	1858	Mexico
vittata Baly	1858	
chalcospila Stal	1858	Mexico
decemlineata (Say)	1824	Mexico, U.S.A.
multilineata (Stal)	1859	
multitaeniata (Stal)	1859	
intermedia Tower	1906	
oblongata Tower	1906	
rubicunda Tower	1906	
defecta (Stal)	1859	Mexico, U.S.A.
dilecta (Stal)	1860	Mexico
patruelis Sturm	1843	
distinguenda Jacoby	1877	Guatemala, Nicaragua
dohrini Jacoby	1883	Mexico
evanescens Stal	1860	Guatemala, Nicaragua, Costa Rica
flavitarsis (Guer.-Meneville)	1855	Mexico
signatipennis Baly	1858	
nitidicollis Stal	1860	
flavopustulata Stal	1860	Guatemala
fraudulenta Kirsch	1874	Peru
guatemalalensis Cockerelle	1920	Guatemala
haldemani (Rogers)	1856	Mexico, U.S.A.
violacea Sturm	1843	
libatrix (Suffrian)	1858	
violascens (Stal)	1858	
litigiosa (Suffrian)	1858	
chlorizans (Suffrian)	1858	
heydeni Stal	1858	Mexico
hogei Jacoby	1883	Mexico
juncta (Germar)	1824	U.S.A.
kirschi Baly	1879	Peru
lacerata Stal	1858	Mexico
hopfneri Dejean	1836	
lineolata (Stal)	1863	Mexico, U.S.A.
melanothorax (Stal)	1859	Mexico, U.S.A.
novemlineata Stal	1860	Mexico
obliterata (Chevrolat)	1833	Mexico
subnotara (Stal)	1858	

paraguensis Jacoby	1903	Paraguay
peninsularis (Horn)	1894	Mexico, U.S.A.
porosa Baly	1859	Brazil
pudica Stal	1860	Mexico
rubiginosa (Rogers)	1856	Mexico, U.S.A.
similis Bowditch	1911	Mexico
stali Jacoby	1883	Mexico
texana Schaeffer	1961	U.S.A.
tlascalana Stal	1858	Mexico, U.S.A.
dahlbomi (Stal)	1859	
tumamoca Tower	1918	U.S.A.
typographica Jacoby	1891	Mexico
undecemlineata (Stal)	1859	Mexico
diversa Tower	1906	
angustovittata Jacoby	1891	
signaticollis (Stal)	1859	
virgulata Archard	1923	Mexico
vittipennis Bowditch	1911	Peru
zetterstedti (Stal)	1859	Mexico

This study includes the 31 species of *Leptinotarsa* in North America including Mexico. In addition to descriptions of each species, information on the host plants, biology, and other interesting information is provided. Identification keys for species of *Leptinotarsa* as well as illustrations of both elytra and genitalia are provided to assist in identification. Historical information on the genus is provided as well as information on the interesting plant family Solanaceae including the potato plant. Illustrations of some of the major host plants are also provided.

This handbook is designed to give the reader more than just the usual information that is included in a taxonomic revision; it is hoped it provides an overview of the species of a very interesting genus of beetles and their close association to their host plants.

CHAPTER 2

PLANTS AND HOST SPECIFICITY

The family Chrysomelidae is well known for having numerous pest species most of which are host specific to certain plants. No green plant escapes from destruction by species of this family. They feed on leaves, stems, and roots. One of the lacunae in our knowledge of the family is a lack of host plant records. According to Jolivet and Petitpierre (1976) only 15 to 20% of the plant hosts of the 35,000 species have been noted leaving 80% of the host plants unknown. Chrysomelids are highly evolved phytophagous beetles so one might expect more host plant records for the family. Wilcox (1979) has provided numerous host plant records for the Northeastern North America species.

Jolivet and Petitpierre (1976) have done considerable research on the host plant associations of chrysomelids, especially the subfamily Chrysomelinae of which *Leptinotarsa* is a member. Jolivet (1986) categorizes types of host plant relationships according to the specific ways in which insects select hosts. The majority of chrysomelids are either monophagous or oligophagous, that is, they feed on a single plant species or at most, species of a few genera. Jolivet's classification of host plant relationship for phytophagous insects consists of four major categories, so called "normal host plants" as follows: monophagy, oligophagy, polyphagy, and pantophagy. Monophagy is subdivided into other categories depending on the number of host plants for the insect species. In first degree monophagy the larvae and adults feed only on one species of plant. Second degree monophagy occurs when the larva or adults feed on several species in the same genus. Third degree monophagy is applied when the larvae or adults feed on all the plant species in a given plant genus. Oligophagous insects feed on plants belonging to several different genera. Again, this category can be further subdivided. First degree oligophagy is the term used when larva and adults feed on related plants belonging to several genera of the same family. Second degree oligophagy is when larvae and adults feed on various genera belonging to related plants of the same order. Finally, third degree oligophagy applies when larva and adults feed on plants of a variety of genera in different plant orders but still in related groups. There are three other categories that Jolivet has established for oligophagous insects, they are: combined oligophagy, when the insect feeds on various genera of a single plant family plus one or more species of plants in a different, unrelated plant family; disjunctive oligophagy is when the insect lives on a limited number of unrelated plants of different orders; and finally, temporary oligophagy occurs when the insect feeds on different and unrelated plants during its larval and adult stages.

Polyphagous insects feed on a great number of plants belonging to unrelated genera in different orders, and pantophagous insects feed on almost all the green flowering plants.

Another term, xenophagy, which Jolivet describes, is the drastic change of the food habits of an animal brought about by extreme habitat stresses. This condition is poorly known but it is mentioned here since it is well known that *Leptinotarsa decemlineata* did indeed change its host plant drastically from Buffalo-bur to the cultivated potato.

Jolivet refers to *Leptinotarsa decemlineata* as an example of restricted oligophagy, that is, feeding on plants of several different genera in the same family.

The majority of work on the genus *Leptinotarsa* and its host plant associations has been done by Ting H. Hsiao of Utah State University. Hsiao and Hsiao (1983) have established the host plants and food habits of 20 species of *Leptinotarsa*, or 50% of the known species, making *Leptinotarsa* one of the better known genera in terms of its host plants. All of the species of *Leptinotarsa* are specialized feeders. Hsiao has found that their host plant associations fall in three plant families. Ten species feed on species of Solanaceae, one on Zygophyllaceae and nine on Compositae. Hsiao has also investigated the host specificity of the solanaceous feeding *Leptinotarsa* (Hsiao 1974) with both field observations and rearing experiments. He has examined the growth and development of eight species of solanaceous-feeding *Leptinotarsa*. His studies indicate that a high degree of host plant specificity is exhibited by *Leptinotarsa* and this in turn reflects a very close adaptation between a number of species of *Leptinotarsa* and the plant family Solanaceae. The analysis of the relationship between *Leptinotarsa* species and their solanaceous host reveals ecological, physiological, and behavioral adaptations (Table IV).

The known life cycles of most *Leptinotarsa* are similar. Eggs are deposited on the foliage; larvae and adults feed on the same host plant, and the pupal stage is in the soil not far from the host plant. There are one or two generations per year according to their geographic location. Depending upon the locations of the species there are at least two strategies of seasonal adaptations for solanaceous feeding *Leptinotarsa*. In the neotropical areas moisture is the most important factor; in the temperate areas of the nearctic region photoperiod is an important factor according to Hsiao (1981). In the warm tropics, as the dry season approaches, the adult beetles enter diapause and wait for the onset of the summer rains when plant growth returns and there is adequate vegetation for feeding. In the temperate zones, as the photoperiod is reduced in the late summer, the adults enter diapause. This ensures the beetle will survive the winter.

TABLE IV. Species of *Leptinotarsa* in the United States & recorded host plants.

SPECIES	LOCALITY	RECORDED HOST PLANTS
L. behrensi	Arizona California Mexico	*Montanoa leucantha*
Present occurence of *L. behrensi* in the United States is doubtful; records for Arizona and California are by state only. These records are over thirty years old.		
L. decemlineata	Widespread USA Mexico	Many recorded host plants including: *Solanum rostratum* *S. tuberosum* and *S. dulcamara* *S. elaeagnifolium* *S. melongena* and *S. triquetrum* *S. angustifolium* *Physalis sp.* and *Lycopersicon esculentum*
L. defecta	Texas Mexico	*Solanum elaegnifolium* *S. tridynamum*
L. haldemani	S. Arizona S. Oklahoma Texas Mexico	*Solanum doughasii* *S. nigrum* *Lycopersicum esculentum* *Physalis viscosa* and *P. acutifolia*
L. juncta	Southeastern USA, extends to North Central and Northeastern states	*Solanum carolinense*

L. lineolata	S. Arizona S. New Mexico S. Texas Mexico	*Hymenoclea monogyra*
L. peninsularis	Arizona Mexico	*Kallstraemia gradiflora*
L. rubiginosa	Arizona New Mexico Texas Mexico	*Physalis sp.* *Solanum pubescens*
L. texana	Texas Mexico	*Solanum elaeagnifolium*
L. tlascalana	Texas Mexico	*Kallstraemia sp.*

The presence of *L. tlascalana* in the United States is doubtful, only US records are from Brownsville, Texas in 1930.

L. tumomoca	Arizona	*Physalis acutifolia*

Total USA species = 11 - 2 (doubtful occurence) = 9 U.S.A. species.

Feeding behavior on the non-solanaceous plants has not been examined to any great extent but host records are increasing. During 1971 Arnett and Jacques (unpublished data) spent the summer in southern Arizona examining the life cycle of *Leptinotarsa lineolata* which feeds exclusively on burro bush, *Hymenoclea monogyra*, a composite and a common plant in the sandy washes in the southwestern United States. The entire life cycle occurs on the plant. Eggs are deposited on the succulent linear-filiform leaves. The larvae hatch in 7 days and feed on the host plant through 4 instars. A prepupal and pupa stage occurs in the soil and transformation to adult occurs in 10 days. What proved to be quite interesting with this species were our attempts to rear this insect on other plants. All attempts failed due to a very simple problem that the larvae of these beetles have. They are unable to cling to other plant leaves because their small larval legs are set close together and adapted to the long slender needlelike leaves of their host plant. This clearly shows a very close adaptation to this plant, a good example of a behavioral and morphological adaptation on the part of *Leptinotarsa lineolata* to feeding on this single species of Compositae. More data will be needed to understand all of the *Leptinotarsa* but these are increasing and as interest in the group continues more information of this type will be gathered.

A list of the major host plants of *Leptinotarsa* follows. The three plant families and the major host species that the *Leptinotarsa* feed on are described and illustrated here.

Hymenoclea monogyra Torre & Gray (Fig. 1)

FAMILY: Compositae

COMMON NAME: Burro-bush

DESCRIPTION: Perennial, broomlike shrub, densely branched, ranging to 6 feet high. Leaves alternate, linear-filiform and entire, or pinnately parted into a few linear-filiform lobes; heads small, those of both sexes usually intermixed in the same leaf axils.

DISTRIBUTION: Locally frequent in sandy desert washes forming thickets. Found in Arizona, New Mexico, southern Texas, portions of California, Baja California, Sonora, and Chihuahua.

HOST PLANT FOR: *Leptinotarsa lineolata*

Kallstroemia grandiflora Torre (Fig. 2)

FAMILY: Zygophyllaceae

COMMON NAMES: Orange Caltrop, Mexican poppy, summer poppy, Arizona poppy.

DESCRIPTION: Annual plant reproducing by seeds, covered with long rough yellowish hairs. This erect plant has stiff hairy stems, branching from the base; stems up to 2 feet long. Leaves opposite, up to 3 inches in length and divided into 5 to 7 pairs of smooth margined hairy leaflets. Flowers large, with deep orange petals from 2/3 to 1-1/4 inches long. Flowers are solitary on slender stalks. Seedpods are green and pearshaped.

DISTRIBUTION: Native plant common in sandy or gravelly soil on mesas, washes, roadsides, and bottom lands in southern and central Arizona. Plants found up to 5,000 feet. Flowers from February to September.

HOST PLANT FOR: *Leptinotarsa peninsularis*

Fig. 1. *Hymenoclea monogyra* Torre & Gray.

Fig. 2. *Kallstroemia grandiflora* Torre.

Physalis acutifolia Gray (Fig.3)

FAMILY: Solanaceae

COMMON NAME: Wright Groundcherry

DESCRIPTION: Annual weed that reproduces by seed. Stems coarse, spreading up to 6 feet. Leaves alternate and vary widely, may be oblong, lanceshaped, or eggshaped, and pointed at the tips. Flowers with a large yellow eye and purplish anthers; flowers are wheelshaped. Flowers may appear in any of the leaf axiles. The calyx is persistent, enlarged, hangs downward, and has the appearance of a "Chinese lantern." The seedpod is berrylike and contains many seeds.

DISTRIBUTION: Wright groundcherry is a native weed of southern and central Arizona. It is a serious pest of irrigated valleys in the state. In addition this plant is found in open ranges, roadside ditches, orchards, and pastures up to 4,000 feet elevation. Flowering occurs from April to November.

HOST PLANT FOR: *Leptinotarsa tumomoca*

Solanum carolinense (Fig. 4)

FAMILY: Solanaceae

COMMON NAME: Horsenettle

DESCRIPTION: Perennial plant developing from creeping roots. Leaves slightly lobed, often oak-like in appearance. Flowers white or lavender with a yellow center, fruit is a yellow berry. Seeds irregular. A spiny plant like buffalo bur, *Solanum rostratum*.

DISTRIBUTION: Widely distributed in the southeastern and eastern states. Found in nearly all habitats especially in the southeast. Spreads aggressively from roots. Considered a noxious weed in many states.

HOST PLANT FOR: *Leptinotarsa juncta* (often called the Horsenettle beetle)

Fig. 3. *Physalis acutifolia* Gray.

Fig. 4. *Solanum carolinense* L.

Solanum elaeagnifolium Cav. (Fig. 5)

FAMILY: Solanaceae

COMMON NAMES: Silver-leafed nightshade; White horsenettle.

DESCRIPTION: Perennial plant similar to horsenettle. Leaves narrower that in horsenettle, entire to coarsely sinuate-dentate, a spiny plant of a silver-white appearance.

DISTRIBUTION: Southwest United States, including Arizona, New Mexico and Texas.

HOST PLANT FOR: *Leptinotarsa texana*, *Leptinotarsa defecta*, and *Leptinotarsa decemlineata*.

Solanum rostratum Dunal. (Fig. 6)

FAMILY: Solanaceae

COMMON NAME: Buffalo Bur

DESCRIPTION: Annual plant; very spiny; leaf segments broad, obtuse; flower with yellow corolla; spines very sharp, straw colored. Leaves pinnatifid or bipinnatifid. Fruit a dry berry enclosed in a spiny covering.

DISTRIBUTION: Southwestern United States, found in uncultivated soil, feed lots, roadside ditches, overgrazed pastures, and about buildings. Considered a pest of range land.

HOST PLANT FOR: *Leptinotarsa decemlineata* (Colorado potato beetle). Considered the original host plant for the Colorado potato beetle.

Fig. 5. *Solanum elaeagnifolium* Cav.

Fig. 6. *Solanum rostratum* Dunal.

CHAPTER 3

EARLY HISTORY OF THE POTATO BEETLE

Some interesting history surrounds the Potato beetles, especially *Leptinotarsa decemlineata*, the Colorado potato beetle, and its rise to fame. This chapter deals with the interesting history of the Colorado potato beetle. In addition, a history of the potato, *Solanum tuberosum* and its fateful ''collision'' with *Leptinotarsa decemlineata* is reviewed.

That Fateful Collision—Potato and Potato Beetle

There are two stores to be told; the first is the interesting history of the potato; the second is the history of *Leptinotarsa decemlineata* (Say) and how this rather obscure insect rose to fame after its fateful meeting with the potato plant.

The story and history of the cultivated potato, *Solanum tuberosum* Linnaeus has been reviewed a number of times in books or portions of books on the subject by Salaman (1949), Heiser (1969), Dodge (1970), and most recently by Klein (1987).

The cultivated potato belongs to the interesting family of plants Solanaceae or Nightshade family. Wild potatoes grow in the southwestern United States, Mexico, Central America, and portions of western South America from Venezuela south through Argentina. In South America the potato is found mainly at higher altitudes, and it is this area that is believed to be the origin of the cultivated potato. It is the underground stem or tuber that is the edible part of the vegetable. The potato, or white potato as it is popularly called, is but one of only six of the 2,000 species of *Solanum* that form tubers. Other very popular edible nightshades are tomatoes, eggplants, and bell peppers.

Potatoes are still a basic food source for the Indians of the Peruvian and Bolivian highlands where the altitudes are too high and too cool to support the growth of corn. Potatoes have been cultivated in these areas for thousands of years. The edible tuber is a good source of starch with little fat and protein. Vitamin C and B-vitamins are found just beneath the skin. A diet of potatoes alone is not adequate for life; the essential vitamins and proteins are lacking. By the tenth century, all highland Indians were under the rule of the Inca tribes from Bolivia who ate the potatoes along with both beans and corn.

The Spanish first saw the potato in 1532 and in 1533 Pedro Cieze de Leon reported on this so called "earth nut" which is boiled and is as tender as a cooked chestnut but has no more skin than a truffle. "Like the truffle it is grown in the earth." Gonzalo Jimenez de Quesado called the tubers "earth truffles." How did the plant first get the name potato? The natives called the plant "papas." This was done to distinguish it from the sweet potato, or batattes, which grew in warmer climates. The sweet potato, *Ipomoea batatas*, is a member of the morning glory family, Convolvulaceae. The natives of South America knew the difference between papas and batattes but the Spanish did not recognize the difference so the word potato is a combination of the two Indian words, "papas" and "batattes." The Spanish sent tubers and seeds back to Spain between 1565 and 1570. The exact source of the potato that reached England in 1580 is not known. John Gerard described and illustrated the potato plant in his Herbal of 1597 and called this the "Potatoes from Virginia." Charles d'Ecluse also illustrated the potato in his Rariorum Plantarum Historia in 1601. In his work he referred to the potato as the "peruvian papas." The actual naming of the potato, that is, the scientific name, was made by Gaspar Bauhin in 1590. He called the potato, *Solanum tuberosum esculentum*, a name which was accepted by Carolus Linnaeus, the founder of the binomial system of nomenclature in the eighteenth century. Linnaeus dropped the third name but retained *Solanum tuberosum*.

Potatoes were not well accepted into Europe at first. In fact it was the first part of the seventeenth century before they were a regular part of the European diet. There were a couple of problems. First, they were in a family of plants which included the nightshades, datura, and mandrake, all used in witchcraft as evil and symbols of death. Some noted that they contained poisons and could "excite Venus," an obvious reference to some sexual connection. As late as 1710 William Salmon, an English physician, said that the "English or Irish potato increaseth thy seed and provoketh lust in both sexes." Some felt that since the skin was rough and scabby there was a relationship between the potato and leprosy. At first all of these ideas prevented the potato from being well accepted on the European table.

It was soon realized that potatoes were not "the wicked plant" like its cousins the nightshade, mandrake, and datura and that leprosy and other diseases were not the result of eating this new vegetable from the New World. It made good sense to grow potatoes. They were easily grown with the use of no or few tools. They matured in four months in all types of soil and the production of one acre could feed an entire family and their livestock! The potato was easy to prepare; simply boil and eat. Therefore it was seen as the economic salvation of the landlord and a way to feed peasants. All types of methods were employed to "sell potatoes" to the masses. Frederick Wilhelm, elector of Prussia, ordered the cultivation of potatoes in 1651. His grandson, Frederick William I, went one step further and passed a law stating that anyone who refused to plant and eat potatoes would have their ears and nose cut off; obviously potato eating became the popular thing to do for the Germans. Peter the Great imported potatoes from Germany for his royal table, and Catherine the Great obtained potatoes and ordered their cultivation in eastern Russia. French prisoners of war, returning from Germany after the Seven Year's War with Prussia, introduced the potato into France since they were fed many potatoes while in Prussia. In 1771 the Academy of Science of Besancon offered a prize for the discovery of a new food which could take the place of cereal grains in the event of famine. Antoine-Augustine Paramentier won the award for recommending that the "cartoufle" or "earth nut," later to be called the "pomme de terre" or "earth apple," be the new food of choice.

The general use of potatoes in the British Isles can be traced to a number of occurrences. In 1662, the Royal Society of London met to consider the possibility of planting potatoes throughout England. Charles II suggested a tax on potatoes realizing that they would be good for England and the money good for his majesty. Some of the first cultivation of the potato can be traced back to the estate of Sir Walter Raleigh in Younghal, Ireland. It was here that Thomas Hariot, estate manager, cultivated the plant. This is interesting since in 1680 Robert Morrison, a professor of Botany at Oxford, wrote that the potato was a familiar English garden plant. Planting stock was obtained from Raleigh's estate in Ireland; thus we see the connection between Ireland and the potato or the "Irish potato" for the first time, but certainly not the last. Nine out of ten people asked where potatoes come from and they say Ireland; like Germany and beer; France and pastry; America and apple pie. The potato and Ireland appear to be inseparable. Indeed the potato has played a major role in the history of Ireland and its people.

Ireland was in need of cheap food. The potato was the answer to the problems of feeding the Irish. The Irish were in an economic depression since the time of Henry VIII and his break with the Roman Catholic Church. The Catholics were forced to be obedient to the English Crown and their land confiscated by Protestants. The Irish tried a number of times to revolt against this harsh treatment by England but they met with failure both in the Desmond Revolt of 1568-1582 and the Cromwellian upheaval of 1650. The Penal Laws of 1690 set in motion rules and regulations that deprived the Irish of their civil rights and the right to any land ownership.

Ireland is an ideal place to grow potatoes; the climate is cool and humid, the soil deep and very workable, and the growing season is long. Combine this with ideal westerly winds off the Atlantic ocean and you have a good place to grow potatoes. They were a natural for Ireland, since from 1700 to 1800 the population of Ireland doubled. Many believed it was this extra food from the potato that contributed to this growth in the Irish population. In 1780 it was said that Ireland's food was potatoes and milk for ten months and potatoes and salt for the other two! The typical Irish family ate potato after potato. A man would consume 12 pounds of boiled potatoes a day, his wife 8 and his children 5. Bread was half potato starch; drink was potato beer and potatoes became the main diet for all of Ireland. Rickets were not uncommon in Ireland since a diet of potatoes lacks calcium and Vitamins A and D. Eye problems and other signs of poor nutrition were seen all over Ireland, but still the potato was the best food for these impoverished people. The Irish were also trapped. They could not afford passage to America; they hated England, and her many lands like Australia were far away. The Catholic Church warned them they would lose their soul if they went off to a pagan country.

So it was in 1845 they were a people tied to England, devoted to the Roman Church, and dependent on the potato for survival. It did not last, although it was not England or the Roman Church, but the potato that failed them. On August 23, 1845, the prestigious Gardener's Chronicle headlined: "A fatal malady has broken out amongst the potato crops." This fatal disease was sweeping Europe and it was most severe in Ireland where potato patches were so close together that any disease could easily spread from one patch to the next. It was obvious that Ireland was doomed. All kinds of suggestions were made on how to use the rotting potatoes but the damage was done and over 80% of the crops were destroyed. Many reasons were given for this malady. Since this was 20 years before Robert Koch and Louis Pasteur proved the relationship between disease and microorganisms, supernatural causes were attributed. Some did note the fungus, *Phytophthora infestians*, but most believed that this was due to rotting potatoes. The three years 1845 to 1847 were the worst. The toll on Ireland was evident. Over 3 million people starved to death; corpses began to rot in the streets; the diseases typhus and cholera became epidemic; and child mortality went to over 65%. Children who survived were in many cases retarded. It was discovered many years later that an alkaloid in the blighted potatoes causes birth defects and abortions. The great famine set into motion a mass emigration, especially to America. This actually reduced Ireland's population from 8 million in 1846 to 4 million in 1900. The Irish made up 35% of the immigrants to America. They flooded the port cities of New York and Boston. They met with new hardships in the new land but they were accustomed to this in Ireland so they survived. As for Ireland the population today is about the same as Connecticut, some 3.5 million inhabitants. Ireland was changed forever by the great potato famine. She lost half of her population to America and over 3 million to starvationı The story of the great potato famine is one of the most tragic pages in human history.

Potatoes entered North America in 1613 when English settlers brought them to Bermuda as food for slaves. The first report of the potato in the United States was in Londonderry, New Hampshire, where it was grown as an animal feed. By 1880 gardeners were attempting to develop cultivars resistant to the disease, rot, scab and all other types of infections. By 1900 the science of plant breeding was well developed. By 1910 geneticists were using wild potato plants from which to select ideal characteristics.

The Great Meet

In 1811 Thomas Nuttal, during a western trip, collected a curious insect feeding on Buffalo-bur, *Solanum rostratum*. The insect was again collected by Thomas Say during Major Long's expedition to the Rocky Mountains in 1819-1820. It was described by Thomas Say in 1824 as *Doryphora decemlineata* the generic name later to be changed to *Leptinotarsa*. For the first 35 years of its recognized existence, the species was of little more than taxonomic interest. American farmers were moving west, especially before and after the Civil War. The American farmer carried with him his beloved potato and all seemed well. Then suddenly in 1859 this new beetle began devastating potatoes growing 100 miles west of Omaha, Nebraska. The speed of the new potato beetle's movement was incredible. The beetle crossed the Mississippi River in 1865, reached Ohio in 1869, Maine by 1972. They reached England in 1875 but there they were exterminated. It developed a strong foothold in Europe during the first World War, being seen in Germany as early as 1914.

The reasons for the rapid change from the Buffalo-bur to the cultivated potato has not been fully explained. The beetles feed on both plants and can still be found on both. Obviously the cultivated potato has a much greater range than Buffalo-bur, and therefore, the beetles are most abundant on potatoes.

The name Colorado potato beetle took a few years to get established. Pope (1984) provides a history of its common name. In the early years the beetle was known as the "ten-striped spearman" (Walsh 1863, Riley 1863), the "ten-lined potato beetle" (Fitch 1863), and the "potato bug" or "new potato bug" (Walsh 1865). The addition of "new" was used to distinguish it from *Lema trilineata* (Olivier), a chrysomelid flea beetle known as the three-lined potato beetle. The term "spearman" was the translation of the generic name *Doryphora*.

The name Colorado was not connected to the beetle when Walsh (1865) stated that in 1864 two of his colleagues had seen the beetle in large number in the new territory feeding on its native host plant the buffalo-bur. This evidence convinced him that Colorado was the source of his beetle so the name Colorado began to appear with the insect. Walsh began to use the phrase "new Colorado potato bug" and it was C. V. Riley (1867) who first used the expression Colorado potato beetle. So within 8 years after its discovery feeding on potatoes west of Omaha we have the common name "Colorado potato beetle" which has been used for this species ever since.

An interesting side light to all this name game occurred in the June 9th, 1866 issue of the Prairie Farmer when a correspondent from Warsaw, Illinois suggested that President Andrew Johnson veto the Colorado bug as well as the Colorado bill for statehood. Colorado's bid for statehood was indeed vetoed by President Johnson in 1867 so Colorado had to wait until President Grant became president in 1876 to become a state. The beetle may have been one of the reasons for Colorado's slow entrance into the Union but the movement of the beetle was anything but slow.

In the very early days hand-picking eggs, larvae and adults was the only method of control. In the time between 1880 and 1900 arsenical insecticides, such as Paris Green, lead arsenate, and Bordeaux mixture, were used to control the beetle. Most were applied by hand sprayers and later by wheeled, horsedrawn sprayers. Power sprayers appeared in 1894 using first steam and later gasoline engines. Other insecticides, such as DDT, aldrin, heptachlor, and kielkrin, joined the list of insecticides used to control this pest in post World War II years. In 1953, Hofmaster reported resistance of the beetle to DDT, and in 1958 it was reported resistant to other chlorinated hydrocarbons. Systemic insecticides followed DDT and its relatives as new weapons in the war against the beetle, but these too have had their problems in recent years, especially with all of the environmental concerns and the developing insecticide resistance.

The other species of *Leptinotarsa* do not have the rich history that has been granted *L. decemlineata*. The only beetle which has an interesting recorded background is *L. juncta*. Blatchley (1910) reported the beetle as "scarce" in southern Indiana being found from June 10 to September 21. The recorded host plant is horse-nettle, *Solanum carolinense*. There has always been speculation that this beetle was "driven out" by the Colorado potato beetle but nowhere are there any accounts of this beetle being found in large numbers. It appears the *L. juncta* occurs throughout southeastern United States and north to southern Illinois, Indiana, Ohio, Maryland, and New Jersey. The beetle can still be found in these areas feeding on horse-nettle.

CHAPTER 4

RESEARCH METHODS

Character States

Character states for the tribe, genus, and species of the genus *Leptinotarsa* have been selected for the analysis of the classification of the group. They include a variety of external morphological features such as color, punctation, and measurements. Methods used in making measurements are described below. An analysis of these character states is important in the development of a sound taxonomic classification and for species identification and separation.

The Groundplan Divergence (GPD) method, as discussed by Wagner (1984), seeks to find the simplest phylogenetic explanation of comparative data. It is concerned with amounts, directions, and sequences of the radiating patterns of genetic divergence. These are estimated on the basis of correlations of character states. Classification, the subject of this work, is usually the by-product of GPD but not the primary goal. GPD is not concerned with formulating taxonomic classifications.

In this work the character states that have been selected are not correlated as genetic relationships. A brief word on this method is appropriate since in further investigations of *Leptinotarsa* species this should be done.

Branching sequences and ancestor-descendant relationships are most important in GPD methods. The more parsimonious and compatible the character correlations are, the more probable the resulting phylogenetic tree. Of secondary importance is chronology, geography, and classifications.

In *Leptinotarsa* we see two species which show a very high degree of correlation: *L. taxana* and *L. defecta*. Not only are they very similar in almost every aspect of their morphological features, but they occur in the same rather restricted geographical area of southern Texas. They are closely related, yet they are separate species. No attempt is made to predict an ancestor for the two species, or if one is the ancestor of the other. This may be accomplished if, and only if, we applied all the GPD methods to the entire genus. The lack of specimens of some species, i.e., those occurring in Central and South American, would affect the resulting phylogenetic tree. Ecological adaptations studies would be useful in the determination of polarity of character states that are believed to be evolutionary adjustments to different life styles. With *L. texana* and *L. defecta* there is some interesting ecological correlations since they feed on not only the same family of plants but the same species. The addition of ecological data for all species would be most interesting since a quarter of the species feed on the same host plant family, Solanaceae, and some species such as *L. lineolata* have actually adapted to their specific host plant due to certain morphological structures.

The question therefore arises as to whether there is coevolution of the insect and the plant species. There is of course a great deal of debate over the addition of such data and what role it plays in the development of a phylogenetic tree. According to Wagner (1984) "any trend in structure or function that supposedly perfects the ability of an organism to survive in a special habitat conditions cannot be accepted until its directionality has been determined first on the basis of GPD." So any attempt to indicate a coevolution of *L. lineolata* with its host plant *Hymenoclea* would not be proper until its place on a phylogenetic tree with the rest of the *Leptinotarsa* is established. None the less, ecological data and adaptation are critical in the study of any organism. The lack of host data for the *Leptinotarsa* and most chrysomelids has proven to be a hindrance to this kind of study.

The character states, as indicated in Table V, are for morphological structures. Ecological data, host plant information, geography, life cycle, larval descriptions, will all enhance in the future construction of a phylogenetic tree for all of the *Leptinotarsa* (Table VI).

TABLE V. Character states

CHARACTER	TRIBE	GENUS	SPECIES
Color			
Unicolous			X
Vittate			X
Punctation			X
Coarse			X
Medium			X
Fine			X
Lacking			X

Size		X	X
Mesosternum		X	
Scutellum		X	X
Elytra:			
Shape			X
Color			X
Unicolorous			X
Vittate			X
Non-vittate			X
Punctation			X
Coarse			X
Medium			X
Fine			X
Lacking			X
Procoxal cavities	X		
General:			
Size	X	X	X
Shape	X	X	X
Color	X	X	X
Unicolorous			X
Vittate			X
Head:			
Shape	X		
Color			X
Unicolorous			X
Spotted			X
Punctation			X
Coarse			X
Medium			X
Fine			X
Lacking			X
Interocular distance			X
Width of head			X
Manibles			X
Maxillae	X	X	X
Antennae			X
Thorax:			
Pronotum:			
Color	X	X	X

Legs:			
Color			X
Unicolorous			X
Spotted			X
Teeth		X	
Tarsal Claws	X		
Abdomen			
Color			X
Unicolorous			X
Markings			X
Male Genitalia			
General Shape		X	X
Size			X
Internal Sac			X

TABLE VI. Character states assigned to ecological and geographical data

HOST PLANT
Plant Family
Plant Genus
Plant Species
GEOGRAPHIC DATA
Plant range
Species range
LIFE CYCLE DATA
Diapause
Larval cycle
instars
Pupa
PrePupa
Predators
Parasites
Pathogens

The above information would be of assistance after the GPD method as described by Wagner was applied to establish a phylogeny.

Taxonomic Characters

Generic characters. The generic characters are limited to external morphological features of adults, including the male genitalia. These characters are: open procoxal cavities and simple claws (separate at base, usually divergent) (Fig. 7). Additional characters useful in delimiting genera include apical segment of maxillary palpi (shorter than penultimate, truncate) (Fig. 8) and mesosternum flat. The head is inserted into the prothorax, the 11-segmented antennae with distal segments usually expanded, the prothorax usually broad and convex, and the elytra convex, covering the abdomen, will assist in defining the genus *Leptinotarsà.*

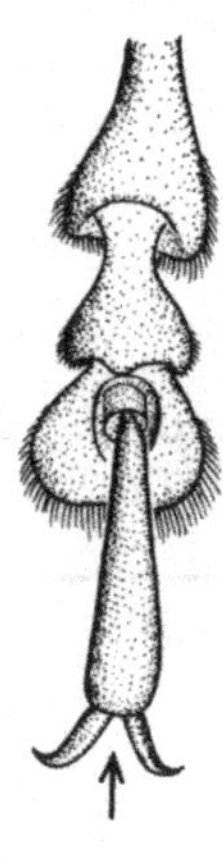

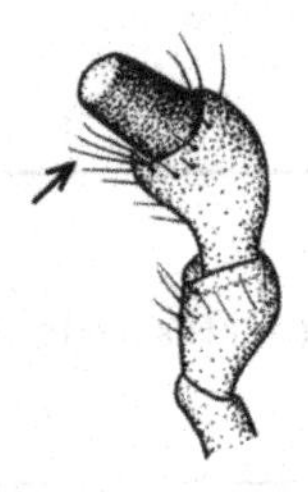

Fig. 8. Apical segments of maxillary palpus of *Leptinotarsa decemlineata* Say.

Fig. 7. Tarsus of *Leptinotarsa decemlineata* Say.

Specific characters. Among the most significant external characters are surface punctations and various elytra markings, including vittae. Body sculpture and coloration are treated in detail for each species. Coloration is useful but in some species there is considerable color variation. Size differences is often useful in separating certain species. The aedeagus is also used for separating species. The characters used for separation of the taxa are listed in Table V.

Terminology followed in genitalia studies is that used by Lindroth and Palmen in Tuxen (1970). The parts of the male genitalia studied are the penis (median lobe) which is the distal (apical) portion, containing the terminal portion and the orifice of the ejaculatory duct, and the tegmen. The term aedeagus is used for the penis and tegmen together.

The penis is tubular (Fig. 9) and contains an internal sac. The internal sac becomes everted into the female vagina during coitus and is thus the functional intromittent organ (Fig. 10).

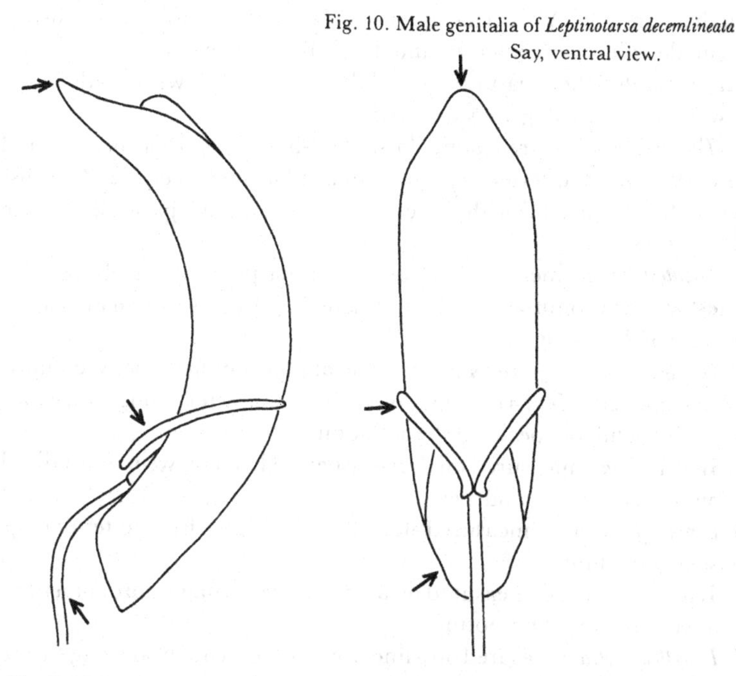

Fig. 10. Male genitalia of *Leptinotarsa decemlineata* Say, ventral view.

Fig. 9. Male genitalia of *Leptinotarsa decemlineata* Say, lateral view.

Species Descriptions

The following notations are given for each of the North American species of *Lepinotarsa*: original description citation; type locality; complete synonymy with complete citation for each synonym; and changes in generic assignment with complete citation. In addition abbreviated citations of useful taxonomic works, such as the Leng and Blackwelder catalogs, follow the other references. Each species redescription contains the following: diagnosis, complete redescription, biology, specimens examined, and discussion. Figures, when available, are included.

The above information would be of assistance after the GPD method as described by Wagner was applied to establish a phylogeny.

Genitalia Preparation. The technique followed here is described in detail by Arnett (1947). Specimens were relaxed by placing them in a beaker of near, but not boiling water, after first removing the labels. They were then removed from the pin in a relaxed condition ready for dissection. An insect pin and forceps were used for removal of the genitalia. The genitalia were placed in a 10% potassium hydroxide solution and heated, this cleared the genitalia of muscle. After clearing, the genitalia were studied and illustrations made. They were later mounted with the specimens.

Measurements. The measurements of the adult beetles were made in the following manner using an ocular grid:

1. *Total length*: measured along dorsal mid-line of head, pronotum, and elytra suture from anterior most point of head to apex of elytra. This distance, although variable within species is useful for the separation of some of the species.

2. *Greatest width*: measured across the widest point of the elytra. The greatest width in conjunction with total length is used to give an indication of the size of the specimens.

3. *Interocular distance*: measured on dorsal aspect of head between most proximal portions of each compound eye. This character, along with width of head, is useful for specific differentiation.

4. *Width of head*: measured on dorsal aspect of head between most distal portions of each compound eye.

5. *Length of pronotum*: measured along dorsal mid-line from anterior edge to base of scutellum.

6. *Width of pronotum*: measured on a line perpendicular to pronotal mid-line at widest point of pronotum.

7. *Length of elytra*: measured on a line from base of scutellum along suture to apex of elytra. Length of elytra along width of elytra is used to give an indication of the size of the specimens.

8. *Width of elytra*: same as number 2 above.

CHAPTER 5

POTATO BEETLE BIOLOGY

The life cycles, or biology, of species of *Leptinotarsa* appear to be quite similar. A general review of the life history, with specific examples follows. After each species description in Chapter 6 additional information can be found on the biologies of *Leptinotarsa* species. Knowledge of the life histories of each species is surprisingly poor but has improved over the past ten years. Once again the Colorado potato beetle leads the group with detailed information on the biology, host plants, pathogens, predators, and parasites. As more and more information on the potential use of these for the control of this pest are acquired more information on the life histories of other *Leptinotarsa* will come to light.

Leptinotarsa species deposit their eggs on the host plant. The eggs are exposed on leaves and stems without any protection. Some species deposit their eggs in clusters in full view on a leaf while others deposit eggs on the underside. Some species such as *Leptinotarsa juncta* deposit only a few eggs in cluster, while *Leptinotarsa lineolata* deposits from 20 to 40, and *Leptinotarsa decemlineata* deposits up to 100 in an irregular mass. Brown, Jermy, and Butt (1980) found the Colorado potato beetle deposits over 3,000 eggs during a female's reproductive lifetime. Eggs hatch in 7 days and the larvae begin to feed immediately on the host plant. Collective feeding by the larvae has been observed for a number of species of *Leptinotarsa*. These observations were made on larvae feeding on the tough leaves of solanaceous plants while no such observations were found with respect to *L. lineolata* which feeds on a composite. For *L. lineolata* the shape of the leaf (linear) was the most critical feature enabling the larva to feed on this host.

The number of instars in usually four. The larval period lasts up to 15 days depending on climatic conditions. Once the larvae stop feeding they enter the soil beneath the plant. There they form a ''prepupa'' stage. This stage is not active; it appears to be a fourth instar larvae except that it is slightly rolled up and does not feed. Pupatation takes place in the soil. The transformation from the larva to adult takes about 10 days making the average life cycle for *Leptinotarsa* beetles about 32 days.

Again, weather and climate will have an effect on the life cycle. As was pointed out in Chapter 2, the Neotropical species differ from the Nearctic species in what triggers adult diapause. In the warm Neotropics it is the amount of moisture, or rather the lack of it, that triggers adults to enter diapause to wait out the dry period. In the Nearactic species such as *L. juncta*, it is the change of photoperiod that triggers adult diapause. As the days get shorter in late summer adults enter diapause, enabling the species to survive the winter season.

The status of host plants and hence, the availability of food, has a major impact on the life cycle of the insect. Hsiao (1986) pointed out that habitat preference is another mode of host adaptation. *Leptinotarsa haldemani* and *L. rubignosa* prefer the shade and are found on solanaceous plants living in the shade, usually under trees or among shrubs. *L. decemlineata*, *L. undecemlineata*, *L. juncta*, *L. defecta*, and *L. tumamoca* are found in open, sunny areas. Some of these species might feed on other plants in the family, but since they prefer either shade or sunny habitats, they remain on plants only in those habitats. The selection of the host plant therefore has as much to do with behavioral modifications as physiological and chemical ones associated with the plant. Even morphological modifications, as was pointed out with the larvae of *L. lineolata*, play an important role in host selection. Unfortunately our present knowledge of *Leptinotarsa* associated with composites is not as advanced as our knowledge of *Leptinotarsa* associated with solanaceous plants.

As has been pointed out briefly, considerable effort has been made to control the Colorado potato beetle and research has advanced our knowledge about this species. In addition to the use of chemical control, there has been considerable effort toward biological control through the use of parasites, pathogens, and predators to control the Colorado potato beetle.

Recently an egg parasite of the Colorado potato beetle has been discovered. This small wasp, *Edovum puttleri* Grissell (Hymenoptera: Eulophidae) was discovered by Ben Puttler in Columbia, South America. Currently tests and evaluation of the eulophid egg parasite are being conducted. Schroder and Athanas (1985) have reviewed early tests and evaluations of this parasite in both lab and field tests. Logan (1983) reported the egg parasites from collections of *L. decemlineata* eggs in Mexico, and on the same field trip, tachnid parasites and several predators were found. The predators included a reduviid bug, nabid bug, and some pentatomids of the genera *Oplomus* and *Stiretrus*.

During the summer of 1971, Jacques and Arnett recorded both egg and larval predators of *L. lineolata* in southern Arizona. Table VII lists the identified egg and larva predators. All were observed and photographed in the field on the host plant (Fig. 11, 12, 13). No predators or parasites of the adults were observed in the field.

Fig. 11. *Calosoma* sp. (carabidae), feeding on a larva of *Leptinotarsa lineolata* (Stal).

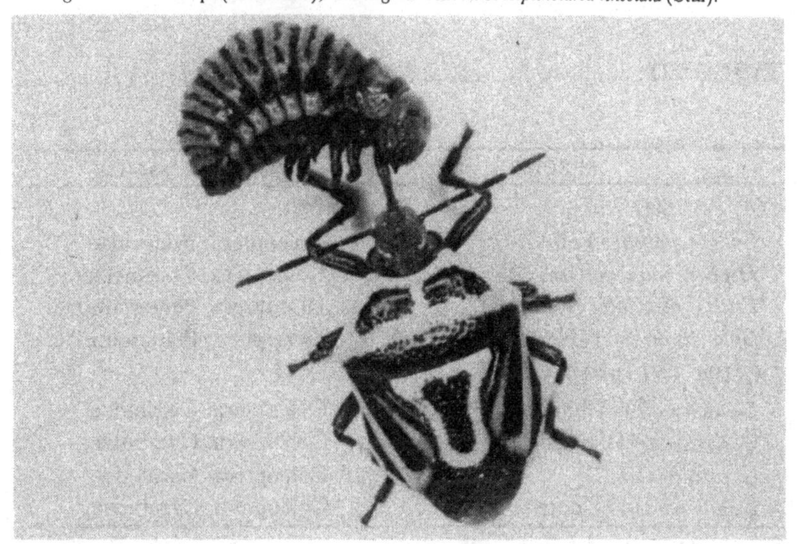

Fig. 12. *Perillus bioculatus* (Fab.) (Hemiptera; Pentatomidae) feeding on the eggs of *Leptinotarsa lineolata* (Stal).

Fig. 13. *Hippodamia convergens* Guer. (Coccinellidae), feeding on the eggs of *Leptinotarsa lineolata* (Stal).

TABLE VII..

SPECIES	ORDER: FAMILY
EGG PREDATORS	
Collops granellus Fall	Coleoptera: Melyridae
Hippodamia convergens Guer.	Coleoptera: Coccinellidae
**Perillus bioculatus* (Fab.)	Hemiptera: Pentatomidae
**Oplomus dichrous* (Herrich-Schaeffe)	Hemiptera: Pentatomidae
LARVAL PREDATORS	
Calosoma affini Chaudoir	Coleoptera: Carabidae
C. angulatum Chevrolat	Coleoptera: Carabidae
C. peregrinator	Coleoptera: Carabidae
C. prominens LeConte	Coleoptera: Carabidae

*egg and larval predators

Numerous carabid predators have been identified by Sorokin (1981) in the Soviet Union. Sorokin identified 11 species. Three species, *Pterostichus cupreus*, *Ophonus rufipes*, and *Broscus cephalotes*, were effective in reducing the populations of the Colorado potato beetle.

Another interesting discovery by Logan *et al.* (1983) is an ectoparasitic mite, *Chrysomelobia labidomerae*, under the elytra of adult beetles. This ectoparasite was found on *L. cacica*, *L. decemlineata*, *L. typographica*, and *L. undecemlineata*.

One species of *Leptinotarsa* may actually be useful in the control of a common, noxious weed in the southeast. Studies by Bailey and Kok (1978) indicate the *L. juncta* may actually be beneficial in the control of the horsenettle, *Solanum carolinense*, in southwest Virginia.

There no doubt remains a considerable amount of information to be gathered on pathogens, predators, and parasites of *Leptinotarsa*.

CHAPTER 6

THE TAXONOMY OF POTATO BEETLES

Introduction

As might be expected of a group on insects with species of economic importance, the genus *Leptinotarsa* has been plagued with name changes.

Nomenclatural History. The name *Leptinotarsa* was first attributed to Chevrolat (1837, in Dejean, Cat. Col. 5:397). The DeJean catalog is a list of names with no descriptions. None of the six names listed beneath *Leptinotarsa* were valid so they are *nomina nuda*, and *Leptinotarsa* is not an available name at that date. *Leptinotarsa* was again listed (1843, in d'Orbigny, Dict. Univ. d'Hist. Nat., 3:656) but was neither described nor had previously described species listed beneath it. The first valid publication of *Leptinotarsa* was by Stal (1858) when he described the genus and included nine species; none was designated as type species. Motschulsky (1860) designated the type-species as "*Lept. Heydenii* Chev.," one of the species included by Stal.

Polygramma was also listed by DeJean (1837 loc. cit.) with citation of four species, all *nomina nuda*. However, listed beneath the *nomem nudum decemlineata* DeJean, as a synonym, was the available name *juncta* Germar (1824); this species serves to validate *Polygramma* with *juncta* as the type-species by monotypy.

In 1974 White and Jacques requested that the International Commission on Zoological Nomenclature use its plenary powers to suppress the generic name *Polygramma* Chevrolat, 1837, for the purposes of the Law of Priority, but not for those of the Law of Homonymy, with *P. juncta* (Germar) as type-species by monotypy on the Official Index of Rejected and Invalid Generic Names in Zoology. In addition it was requested to place the generic name *Leptinotarsa* Stal 1858, (gender: feminine) on the Official List of Generic Names in Zoology with the type-species *L. heydenii* Stal, as designated by Motschulsky, 1860, and to place *Leptinotarsa heydenii* (Stal), 1858, on the Official List of Specific Names in Zoology. This request was based on the fact that the firmly established name of *Leptinotarsa* appears in thousands of publications, especially in the literature of economic entomology, with reference to the Colorado potato beetle, *Leptinotarsa decemlineata* (Say). Changing the name from *Leptinotarsa* back to *Polygramma*, therefore, is not practical.

The International Commission of Zoological Nomenclature, in its Opinion Number 1290 (1985), has conserved *Leptinotarsa* Stal 1858 and has rejected *Polygramma* as requested by White and Jacques (1974).

Additional uses of the name *Leptinotarsa* appear after 1858; Stal (1862-1865) includes all of the species of *Leptinotarsa* and *Myocoryna* in the genus *Chrysomela*. He makes no reference to *Polygramma*. Crotch and Cantab (1873) split Stal's *Chrysomela* into 5 genera; *Labidomera*, *Myocoryna*, *Zygogramma*, *Calligrapha*, and *Chrysomela*. Chapuis (1874), in a key to the *Chrysomelidae*, included the genus *Leptinotarsa* Chevrolat, but not *Myocoryna* or *Polygramma*.

Jacoby (1883) placed *Myocoryna* in synonymy with *Leptinotarsa* and included some species from *Chrysomela* and *Doryphora* in the genus *Leptinotarsa*. Jacoby states that the genus *Leptinotarsa* was erected by Stal. After Jacoby's work, all subsequent citations of *Leptinotarsa* are credited to Stal (1858) not Chevrolat (1836). Jacoby's first treatment of *Leptinotarsa* included 35 species. Jacoby (1891) added 4 species to *Leptinotarsa* and illustrates many of the *Leptinotarsa* in accurate, detailed colored plates. Of the 39 species included in the genus, 4 are recorded exclusively from Central and South America. The remaining species are found in Mexico and the United States.

Linell (1896) published a key to the North America species of *Leptinotarsa* and included 12 species from the United States, all from the southwest except for *L. decemlineata* and *L. juncta*.

Tower (1906) published the results of extensive investigations on the evolution of beetles in the genus *Leptinotarsa*. Knab (1908) presents a discussion of Tower's publication and points out some of the errors. Tower claimed only three species of *Leptinotarsa* were found in the United States when actually there were at least 11 recorded at that time. There are 9 confirmed species found in the United States as reported in this publication. He failed to cite Stal's classic work on the Chrysomelidae of America and his work lacks an index which makes it difficult to use. The work is really a treatise on the Colorado potato beetle, *L. decemlineata*. He spent considerable time with genetic studies on this species and he names 9 varieties. Tower described 5 new species, 4 of which are placed in synonymy in this work. His descriptions were poor and he recorded little more than locality data in his 1907 publication. Tower claimed all the species of

Leptinotarsa feed on plants in the family Solanaceae; this statement is incorrect since the majority of the United States species actually feed on members of the plant family Compositae. According to Hsiao (1986), 25% of the species of *Leptinotarsa* feed on Solanaceae, 5% on the Zygophyllaceae and 70% on the Compositae. Towers chief concern was not taxonomy but evolution and genetics. His work contains some very fine material, but the taxonomic material should be viewed with careful scrutiny.

After the Tower publication there was little published on the genus *Leptinotarsa*. Knab (1908) published notes on *Leptinotarsa undecemlineata* (Stal) and cleared up the confusion between *L. texana* Schaeffer and *L. defecta* (Stal). *L. texana* Schaeffer did not receive full species status until Brown (1961) reviewed some North American Chrysomelidae.

Six new species of *Leptinotarsa* have been described since 1906 and all but one are retained. In 1911, *L. similis* was described by Bowditch from Mexico, *L. tumomoca* was described by Tower from Arizona in 1918, *L. boucardi* and *L. virgulata* were both described from Mexico by Achard in 1923, *L. texana* Schaeffer received full species status in 1961. The final species named *L. collinsi* described by Wilcox in 1972 from Arizona is the species that is not retained. Examination of the type reveals this species not to be in the genus *Leptinotarsa*, but in the genus *Calligrapha*.

The tribal classification for *Leptinotarsa* has changed at least three times. Blatchley (1910), Weise (1916), Leng (1920), and Blackwelder (1946) placed *Leptinotarsa* in the tribe Chrysomelini. Dillon and Dillon (1961) place the *Leptinotarsa* in the tribe Sygogrammini. Wilcox (1972) and Arnett (1963, 1985) place the *Leptinotarsa* in the tribe Doryphorini.

Key to the North American Genera of Doryphorini

(modified from Wilcox 1972)

Members of the Tribe Doryphorini have front coxal cavities open behind and simple tarsal claws.

1. Claws, parallel and contiguous at base; elytra vittate or spotted .... *Zygogramma* Chevrolat

—Claws divergent or at least separated at base (Fig. 7) 2

2(1). Apical segment of maxillary palpi as long as or longer than the preceding, truncate (Fig. 8) 3

—Apical segment of maxillary palpi as long as or longer than the preceding, dilated, truncate 4

3(2). Front femur of male strongly toothed, normal in female; mesosternum forming a blunt tubercle between the middle coxae, raised above the level of the prosternum .... *Labidomera* Chevrolat

—Front femur of male and female normal; mesosternum not raised above the level of the prosternum .... *Leptinotarsa* Stal

4(2). Elytra with spots or vittate delimited by punctures; pronotum not thickened at sides, without distinct longitudinal impressions .... *Calligrapha* Chevrolat

—Elytra entirely dark or dark with pale lateral margins; elytra punctures usually in irregular rows; sides of pronotum thickened, the thickened portion separated from disc by a longitudinal impression .... *Chrysolina* Motschulsky

Genus *LEPTINOTARSA* Stal 1858

Leptinotarsa Chevrolat in Dejean, Cat. Col, 1936:421. (*Nom. nud.*): Sturm, 1843:286. (*Nom. nud.*); Stal, 1858. Ofv. Svenska Vet.-Akad. Forh., 15:475; Motschulsky, 1860:181; Chapuis, 1874:368; Jacoby, 1883:227; Linell, 1896:195; Blatchley, 1910:1152; Bradley, 1930:253; Wilcox, 1954:413; Arnett, 1963:909.

Type species of Genus.—Leptinotarsa heydeni Stal 1858:475; type of subsequent designation, Motschulsky, 1860:182.

Myocoryna Dejean, Cat. Col., 1836:428. (*Nom. nud.*); Stal 1859. Ofv. Svenska Vet.-Akad. Forh., 16:316. Crotch, 1873:46.

Type species of Genus.—Myocoryna juncta (Germar) 1824:590; here designated.

Polygramma Chevrolat in Dejean, Cat. Col., 1836:421. Motschulsky 1860. Schrensk's Reisen Amurl., 2:181.

Type species of Genus.—Polygramma juncta (Germar) 1824:590; type by subsequent designation in Motschulsky 1860:181.

Open procoxal cavities, simple tarsal claws divergent, separated at base, maxillary palpi with apical segment shorter than preceding, truncate apical segment, third tarsal segment entire or scarcely emarginate, mesosternum not raised above level of prosternum are characters which will separate this genus from all other genera of Chrysomelinae in North America.

Description of Genus. These are medium to large beetles, ranging from 6.0

mm to 18.0 mm long, 5.0 mm to 11.5 mm wide, broadly oval, convex, variable coloration including vittate, unicolorous, spotted, metallic, and non-metallic species.

Head: inserted into the prothorax to the eyes, deflexed; antennae moderately long, 11-segmented, third segment twice the length of second, clavate; labrum variable; mandibles short, stout, apices blunt; maxillary palpi 4-segmented, apical segment shorter than preceding, truncate, subquadrangular or dilated; eyes feebly emarginate, moderate in size, lateral, elongate.

Thorax: pronotum broader than head, lateral margins slightly curved, anterior-lateral angles distinct, rounded or pointed; punctation fine to course, even or unevenly distributed, sparse to dense; procoxal cavities open; prosternum narrow between coxae; scutellum triangular.

Legs: moderate in size; procoxae transversely oval, separate; femur slightly swollen; tibiae broader towards apex; third tarsal segment entire; claws simple, separate at base.

Elytra: oval to oblong oval; wider at base than pronotum; punctation variable; coloration variable.

Abdomen: five visible sterna, free; fine pubescense; coloration variable.

Male genitalia: size of aedeagus variable, penis curved, cylindrical; tegmen closer to base.

Sexual dimorphism: females slightly larger than males in some species; gravid females much more robust than males.

Larva: three to four instars; cyphosomatic; 2.5 - 9.0 mm long; color variable but usually creamy white in the first two instars.

Prepupa: a quiescent instar between the last larval instar and the pupa; resembles the last larval instar.

Pupa: execrate; length 10 mm (average); creamy white; pigmentation in spiracle area.

Female genitalia were not studied.

Distribution.—This genus, limited to the Western Hemisphere, extends from the southwestern United States to Peru and Brazil. Its greatest development is in Mexico where 29 species are recorded, 9 species are found in the United States. Species in the United States are found mainly in Texas, New Mexico, Arizona and Southern California, with *L. juncta* (Germar) in the southeastern states and *L. decemlineata* (Say) found throughout the United States, Mexico, southern Canada, and has also spread to Europe and parts of Asia.

Chronological list of names used for the species of North America:

1824. *Chrysomela juncta* Germar
1824. *Chrysomela decemlineata* Say
1833. *Doryphora obliterata* Chevrolat
1836. *Leptinotarsa hopfneri* Dejean
1854. *Doryphora rubiginosa* Rogers
1855. *Polyspila flavitarsis* Guerin-Meneville
1856. *Doryphora haldemani* Rogers
1858. *Leptinotarsa cacica* Stal
1858. *Leptinotarsa calceata* Stal
1858. *Leptinotarsa vittata* Baly
1858. *Leptinotarsa chalcospila* Stal
1858. *Doryphora cholrizans* Suffrian
1858. *Leptinotarsa signatipennis* Baly
1858. *Leptinotarsa heydeni* Stal
1858. *Leptinotarsa lacerata* Stal
1858. *Doryphora libatrix* Suffrian
1858. *Doryphora chlorizans* Suffrian
1858. *Leptinotarsa signatipennis* Baly
1858. *Leptinotarsa tlascalana* Stal
1859. *Myocoryna dahlbomi* Stal
1859. *Myocoryna multilineata* Stal
1859. *Myocoryna defecta* Stal
1859. *Myocoryna violascens* Stal
1859. *Myocoryna melanothorax* Stal
1859. *Myocoryna signaticollis* Stal
1859. *Myocoryna undecemlineata* Stal
1859. *Leptinotarsa zetterstedti* Stal
1860. *Leptinotarsa dilecta* Stal
1860. *Leptinotarsa nitidicollis* Stal
1860. *Leptinotarsa novemlineata* Stal
1860. *Leptinotarsa pudica* Stal
1863. *Chrysomela lineolata* Stal
1877. *Leptinotarsa behrensi* von Harold
1883. *Leptinotarsa modesta* Jacoby
1883. *Leptinotarsa puncticollis* Jacoby
1883. *Leptinotarsa dohrini* Jacoby
1883. *Leptinotarsa hogei* Jacoby
1883. *Leptinotarsa stali* Jacoby
1891. *Leptinotarsa typographica* Jacoby

1891. *Leptinotarsa angustovittata* Jacoby
1894. *Myocoryna peninsularis* Horn
1906. *Leptinotarsa decemlineata* var. *texana* Schaeffer
1906. *Leptinotarsa diversa* Tower
1906. *Leptinotarsa intermedia* Tower
1906. *Leptinotarsa oblongata* Tower
1906. *Leptinotarsa rubicunda* Tower
1906. *Leptinotarsa decemlineata* var. *albida* Tower
1906. *Leptinotarsa decemlineata* var. *pallida* Tower
1906. *Leptinotarsa decemlineata* var. *minuta* Tower
1906. *Leptinotarsa decemlineata* var. *obscurata* Tower
1906. *Leptinotarsa decemlineata* var. *tortuosa* Tower
1906. *Leptinotarsa decemlineata* var. *defectopunctata* Tower
1906. *Leptinotarsa decemlineata* var. *melanicum* Tower
1906. *Leptinotarsa decemlineata* var. *rubrivittata* Tower
1906. *Leptinotarsa decemlineata* var. *immaculothorax* Tower
1911. *Leptinotarsa similis* Bowditch
1918. *Leptinotarsa tumomoca* Tower
1923. *Leptinotarsa boucardi* Achard
1923. *Leptinotarsa virgulata* Achard
1961. *Leptinotarsa texana* Schaeffer

Key to the species of *Leptinotarsa* of North America

1. Head, thorax, and elytra unicolorous, without markings or vittae . . . 2
—Thorax and/or elytra with some type of maculation or vittae 5
2(1). Elytra aeneous, blue, or green 3
—Elytra pale yellow or orange-red 4
3(2). Head, thorax, and elytra aeneous to blue, head and thorax with dense, fine punctation, elytra punctation course, dense (Arizona, Mexico. . . . . *behrensi* Harold
—Head, thorax, and elytra metallic green, blue or violet, head thorax, and elytra with fine sparse punctation (Texas, New Mexico, Arizona, Mexico) . . . . *haldemani* (Rogers)
4(2). Head, thorax, and scutellum metallic blue, elytra pale yellow with metallic blue sutural margin (Mexico). . . . . *cacica* Stal
—Head, thorax, and elytra orange-red, scutellum black (Arizona, Mexico). . . . . *rubiginosa* (Rogers)

5(1). Thorax and elytra with maculations, vittae, or darkened punctation 6
—Thorax immaculate, unicolorous, elytra with vittae or maculations . 11
6(5). Each elytron with 2 incomplete, greatly reduced vittae present, elytra punctation incomplete, in irregular rows from base to apex (Texas, Mexico) . . . . *defecta* (Stal)
—Each elytron with at 4 complete vittae extending from base to apex of elytra, punctation variable 7
7(6). Each elytron with 4 vittae, vitta 1 shorter than other 3, vittae 2, 3, and 4 join at apex (Texas) . . . . *texana* Schaeffer
—Each elytron with 5 vittae, vitta 5 parallel to lateral margin of elytra 8
8(7). Small, total length 7.3-8.1 mm; greatest width 5.2-5.8 mm, vitta 2, 3, and 4 join at apex of elytron (Arizona). . . . . *tumamoca* Tower
—Large, total length 8.3-13.1 mm; greatest width 5.3-8.0 mm, vittae not joining at apex as above 9
9(8). Head, and thorax with various markings, ventral surface and legs unicolorous, black (Mexico) . . . . *undecemlineata* (Stal)
—Head and thorax with various markings, ventral surface and legs multicolored 10
10(9). Elytra punctation in regular rows from base to apex of elytra, black spot on outer margin of the femur (Southeast U.S.) . . . . *juncta* (Germar)
—Elytra punctation irregular, not forming regular rows, no black spot on legs (widespread) . . . . *decemlineata* (Say)
11(5). Large, total length 12.0-15.5 mm; greatest width 8.6-11.0 mm; elytra base color either blue-black with flavous markings or chestnut brown with black spots 12
—Total length under 12.0 mm; greatest width under 8.6 mm; elytra neither blue-black or chestnut brown; elytra base color either pale yellow, brown, or reddish 14
12(11). Elytra chestnut brown with 7 to 10 dark black spots on each elytron (Mexico) . . . . *chalcospila* Stal
—Elytra dark blue to black with flavous markings 13
13(12). Elytra dark blue to black with 3 flavous cross bands and a flavous spot at the apex of each elytron (Mexico) . . . . *lacerata* Stal
—Elytra dark blue, each elytron with at least 6 irregular flavous spots, the middle spot near sutural margin sometimes confluent with the elongated spot near the lateral margin forming a transverse band (Mexico) . . . . *heydeni* Stal
14(13). Elytra with distinct, complete, non-interrupted vittae, no other

elytra markings present. 15
—Elytra with interrupted vittae, vittae and spot combination or other markings 19
15(14). Thorax unicolorous, bronze to metallic green or blue. 16
—Thorax unicolorous, brown to black 17
16(15). Thorax bronze to coppery-green, each elytron with 4 distinct vittae, elytra punctation course, dense (Mexico). . . . .*novemlineata* Stal
—Thorax metallic blue or green, each elytron with 4 distinct vittae, elytra punctation sparse . . . .*calceata* Stal
17(15). Total length 8.4-10.2 mm; greatest width 6.5-7.7 mm; head, pronotum, legs and abdomen unicolorous, black; elytra pale yellow with 4 distinct vittae and a faint fifth parallel to lateral margin (Mexico) . . . .*melanothorax* (Stal)
—Total length 6.5-7.5 mm; greatest width 4.5-5.5 mm; head, pronotum, legs, and abdomen unicolorous, reddish brown to black; elytra flavous or brown, each elytron with 2 or 3 vittae. 18
18(17). Head and pronotum brown to rufescent, elytra flavous, and each elytron with 3 brown vittae, vittae 2 and 3 join at apex, lateral margin brown (Arizona, Mexico) . . . .*peninsularis* (Horn)
—Head and pronotum dark brown to black, elytron dark brown with 2 flavous vittae joining at apex (Texas, Mexico) . . . . .*tlascalana* (Stal)
19(14). Thorax unicolorous, rufescent, or reddish-yellow 20
—Thorax unicolorous, bronze, cuperous, metallic blue or green 23
20(19). Elytra either pale-yellow or reddish-yellow, elytra with vittae, vitta 1 complete, other vittae subdivided in apex area of elytra 21
—Elytra rufescent, non-vittate, with yellow maculation in half moon shape outlined in black, yellow mark along lateral margin with 4 projections toward sutural margin (Mexico). . . . .*zetterstedti* Stal
21(20). Each elytron with 4 black vittae intersected at the middle by a transverse light brown line, vittae 1, 2, and 5 complete to apex, vittae 3 and 4 subdivided into fragments at apex (Mexico)
. . . .*boucardi* Achard
—Each elytron with 4 vittae at base, vittae 3 and 4 meeting near middle of elytron, vitta 1 and 2 complete, apex of elytron with spots and fragments of vittae 22
22(21). Pronotum reddish-brown, elytra pale yellow, vittae and spots black (Mexico) . . . .*typographica* Jacoby
—Pronotum reddish-brown, elytra reddish-brown, vittae red (Mexico)*dilecta* Stal
23(19). Elytra non-vittate, pale yellow with 3 to 4 spots on each elytron
. 24

—Elytra vittate, usually pale yellow but not marked with spots as above, with vittae and spots, many spots or other markings 25

24(23). Total length 11.0-12.0 mm; greatest width 7.5-8.0 mm; elytra flavous, each elytron with 3 dark brown irregular spots, punctation course in double irregular rows (Mexico). . . . .*obliterata* (Chevrolat)

—Total length 9.0 mm; greatest width 6.5 mm; elytra flavous, each elytron with 3 to 5 irregular brown spots, punctation course, darkened forming irregular rows (Mexico). . . . .*pudica* Stal

25(24). Elytra without spots exclusively 26

—Elytra testaceous with numerous black spots, spots smaller at lateral and sutural margin, larger and confluent at middle of elytron (Mexico-) . . . .*dohrini* Jacoby

26(25). Sutural margin thick, at least 1 mm wide, dark green to black; elytra pale yellow marked with black or dark green maculation non-vittate 27

—Sutural margin, if present, very thin elytra marked with distinct vittae. 28

27(26). Sutural margin dark green, elytra flavous with a thick dark green mark extending from base of elytron to apex (Mexico).*hogei* Jacoby

—Sutural margin black, elytra pale yellow, elytra marked with black maculations, large black area at base and transverse band at disk, small black dot at apex of elytron (Mexico). . . . .*stali* Jacoby

28(26). Elytra with 3 or 4 interrupted vittae, vitta thickened at various points 29

—Elytra with vittae and other maculations, only 1 or 2 complete vittae30

29(28). Elytra with 4 interrupted vittae, vittae 3 and 4 join at base, each vittae interrupted at the center of the elytron (Arizona, Mexico) . . . .*lineolata* (Stal)

—Elytra with 3 vittae, vittae 1 and 2 thickened at base, extending to apex, vitta 3 begins at middle of elytron, thickened, vittae join at apex (Mexico) . . . .*flavitarsis* (Guerin-Meneville)

30(28). Elytra with 1 complete vittae adjacent to sutural margin, other vittae short, does not extend beyond base; apex of elytra with fragments of vittae 2, 3, and 4 (Mexico) . . . .*similis* Bowditch

—Elytra with 2 fine vittae, vitta 2 shorter than vitta 1, and joining an oblique mark at apex, elongated comma mark at middle of elytron (Mexico) . . . .*virgulata* Achard

Key to the Species of *Leptinotarsa* of the United States

1. Head, thorax, and elytra unicolorous, without markings or vittae . . .2

—Thorax and/or elytra with some type of maculation or vittae 3

2(1). Head, thorax, and elytra metallic green, blue, or violet; head, thorax, and elytra with fine, sparse punctation. (Texas, New Mexico, Arizona) . . . .*haldemani* (Rogers)

—Head, thorax, and elytra orange-red; scutellum black (Arizona, New Mexico) . . . .*rubiginosa* (Rogers)

3(1). Thorax and elytra with maculations, vittae, or darkened punctation 4

—Thorax immaculate, unicolorous; elytra with vittae or maculations 8

4(3). Each elytron with two incomplete, greatly reduced vittae present; elytra punctation incomplete, in irregular row from base to apex (Texas) . . . .*defecta* Stal

—Each elytron with four complete vittae extending from base to apex of elytra, punctation variable 5

5(4). Each elytron with four vittae, vitta one shorter than other three, vittae 2, 3, and 4 join at apex (Texas) . . . .*texana* Schaeffer

—Each elytron with 5 vittae, vitta 5 parallel to lateral margin of elytra 6

6(5). Small, total length 7.3-8.1 mm; greatest width 5.2-5.8 mm; vitta 2, 3, and 4 join at apex of elytron (Arizona). . . . .*tumamoca* Tower

—Large, total length 8.3-13.1 mm; greatest width 5.3-8.0 mm; vittae not joining at apex as above 7

7(6). Elytra punctation in regular rows from base to apex of elytra; black spot on outer margin of the femur (Southeastern U.S.) . . . .*juncta* (Germar)

—Elytra punctation irregular, not forming regular rows, no black spot on leg (widespread in U.S.) . . . .*decemlineata* (Say)

8(3). Dark markings of elytra consist of unbroken vittae; the humeral vittae may be very short but there are no spots, each elytron with three brown vittae, vittae 2 and 3 join at apex, lateral margin brown (Arizona) . . . .*peninsularis* (Horn)

—Some of the elytra vittae broken, small dark spots also present on elytra; elytra with 4 interrupted vittae, vittae 3 and 4 join at base, each vittae interrupted at the center of the elytron; 7-7.7 mm long (Arizona, New Mexico, Texas) . . . .*lineolata* (Stal)

Descriptions of the Species

Leptinotarsa behrensi Harold

Leptinotarsa behrensi Harold, 1877. Mitt. Munchener Ent. ver. 1:16.
Type locality: California.

Leptinotarsa modesta Jacoby, 1833. Biologia Centrali-Americana, Insecta, Coleoptera, Chrysomelidae. 6(1):229.
(Synonymy, Knab 1908:224.)
Type locality: Guanajuato, Mexico

Leptinotarsa puncticollis Jacoby, 1883. Biologia Centrali-Americana, Insecta, Coleoptera, Chrysomelidae. 6(1):228.
(Synonymy, Knab 1908:224.)
Type locality: Somora and Ventanas, Durango, Mexico

Jacoby, 1891:253; Linell, 1896:196; Leng, 1920:294; Blackwelder 1946:673; Wilcox 1972:9.

The shinny copper-green coloration of the head, pronotum and elytra, dense, fine punctation of head and pronotum and course punctation of elytra will separate this species from all others in North America. (*See back cover, lower right.*)

Head: punctation fine; proximal 6 segments of antennae slender, shinny, distal 5 segments expanded, dull, with fine pubescence; interocular distance 2.0 mm, head width 2.7 mm; eyes feebly emarginate.

Thorax: pronotum length 2.7 mm, width 7.0 mm; pronotum shinny, immaculate; punctation course at lateral margins, fine at disk; anterior-lateral angles bluntly pointed. Scutellum shinny, punctation fine, dense. Elytra broadly oval, convex; punctation irregular, course, dense; pronotum length 10.6 mm, pronotum width 9.6 mm.

Legs: femur and tibia with course punctation; distal part of tibia with fine, thick, golden pubescence; unicolorous, copper-green.

Abdomen: unicolorous, bronze; punctation fine.

Individual variation.—Total length 14.5 ± 1.40 (13.0-17.5); greatest width 10.0 mm ± .85 (9.0-12.0); interocular distance 2.0 mm ± .09 (2.0-2.2); head width 2.7 mm ± .16 (2.5-3.0); pronotum length 2.7 mm ± .32 (2.4-3.1); pronotum width 7.0 mm ± .41 (6.3-7.7); elytra length 11.3 mm ± 1.18 (9.7-13.7); elytra width 10.0 mm ± .85 (9.0-12.0).

The color of the head, pronotum, and elytra will vary. Three specimens studied were blue rather than the usual copper-green color. Two other specimens were brown, but the majority of specimens examined were cooper-green in color.

Specimens examined.—30: ARIZONA: 4, (CASC); MEXICO: CHIHUAHUA: 1, Barranca del Cabre, July 20, 1957 (TAMC); 1, Pinos Altos (USNM); DURANGO: 1, Durango to Pacific, (USNM); JALISCO: 2, Guadalajara, Oct. 20, 1905 (USNM); 3, Colimilla Barranea de Ovlates, Aug. 8, 1958 (AMNH): 1, Rio Verde, July 5, 1953 (AMNH); 1, Guanajuato, (AMNH); 5, 18 mil S. Acatlan, July 23, 1966 (TAMC); SINALOA: 1, (AMNH); 1, (USNM); SONORA: 9, Guirocoba, July 1, 1933 (LACM).

This species is found mainly in northern and central Mexico (Fig. 14, map).

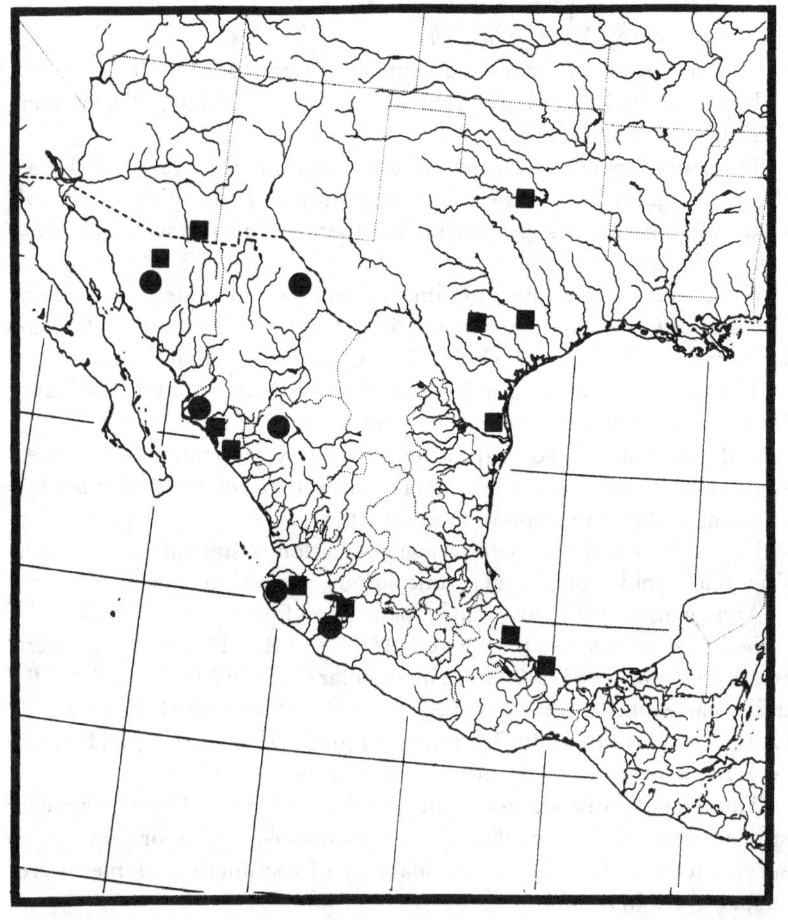

Fig. 14. Map of the distribution of *L. behrensi* Harold (circle), and *L. haldemani* (Rogers) (square).

It is doubtful if *L. behrensi* occurs in the United States. A series of 4 specimens only had the state label Arizona but no recent material with locality details has been collected. This species has been recorded in the Mexican states of Sonora, Chichuacha, Durango, Jalisco, and Sinaloa. Hsiao and Hsiao (1983) reported collecting this beetle on *Montanoa leucantha* (Lag.) DC at Lake Chapala south of Guadalajara in the state of Jalisco, Mexico.

Leptinotarsa haldemani (Rogers)

Leptinotarsa violacea Sturm, 1843. Catalog der Kaefer-Sammlung von Jacob Sturm. 386 pp. (*Nom. Nud.*)
Type locality: Mexico

Doryphora haldemani Rogers, 1854. Proc. Acad. Nat. Sci. Philadelphia. 8:30
Type locality: Fredericksburg, Texas.

Doryphora chlorizans Suffrian, 1858. Stettiner Ent. Zeitung. 19:248. (NEW SYNONYMY.)
Type locality: Mexico

Doryphora libatrix Suffrian, 1858. Stettiner Ent. Zeitung. 19:248. (NEW SYNONYMY.)
Type locality: Mexico

Myocoryna violacscens Stal, 1859. Ofv. Svenska Vet.-Akad. Forh. 16:317. (NEW SYNONYMY.)
Type locality: Mexico

Chrysomela haldemani Stal, 1863. Monographie des Chrysomelides de l'Amerique. 2:167. (New generic assignment.)

Myocoryna haldemani Crotch, 1873. Proc. Acad. Nat. Sci. Philadelphia, 25:47. (New generic assignment.)

Leptinotarsa haldemani Jacoby, 1883. Biologia Centrali-Americana, Insecta, Coleoptera. 6(1):235. (New generic assignment.)

Suffrian 1858:246; Linell 1896:196; Tower 1906:4; Leng 1920:294; Powell 1941:163; Blackwelder 1946:673; Wilcox 1972:9.

The metallic blue or green elytra with fine, sparse punctation, head and thorax metallic green to blue-black, ventral surface and legs black, unicolorous will separate this species from all others in North America. (*See back cover, upper left.*)

Head: unicolorous, metallic green or dark blue-black; punctation fine, sparse; interocular distance 1.8 mm, head width 2.0 mm.

Thorax: unicolorous, metallic green or dark blue-black, same color as head; punctation fine, sparse; anterior-lateral angles rounded; pronotum length 2.5 mm, width 5.0 mm. Scutellum black, smooth, Elytra oval, convex; immaculate, metallic green or blue to violet; punctation fine, forming irregular rows; elytra length 7.3 mm, elytra width 6.9 mm.

Legs: unicolorous, black.

Abdomen: unicolorous, black.

Male genitalia.—Aedeagus cylindrical, arched; 4 times as long as wide; apex of apical region bluntly pointed; basal area flattened, expanded. (Figs. 15 and 16).

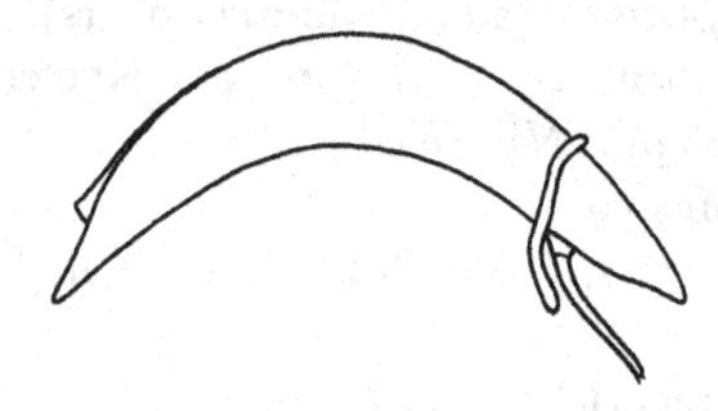

Fig. 15. Male genitalia of *L. haldemani* (Rogers), lateral view.

Fig. 16. Male genitalia of *L. haldemani* (Rogers), ventral view.

Females.—Gravid females larger than males.

Individual variation.—Total length 9.5 mm ± .56 (8.8-11.0); greatest width 6.9 ± .47 (6.2-7.8); interocular distance 1.8 mm ± .12 (1.6-2.0); head width 2.0 mm ± .30 (1.9-2.4); pronotum length 2.5 mm ± .22 (2.1-3.0); pronotum width 5.0 mm ± .35 (4.4-5.8); elytra length 7.3 mm ± .49 (6.5-7.9); elytra width 6.9 mm ± .47 (6.2-7.8).

This species has some interesting color variation. The colors of the elytra vary from metallic green to blue to a blue-black or violet, but this is not geographically correlated.

Specimens examined.—(Fig. 14, map) 149:OKLAHOMA: 4, 2 mi. E. Willis, Lake Texoma, June, 1965 (UCDC); 1, 2 mi. E. Willis, Lake Texoma, June, 1965 (OSUC), TEXAS: 1, (UCDC); 10, (USNM); 1, Brownsville, June 23, 1930 (UCDC); 1, Austin Co., April 1, 1954 (OSUC); 1, San Antonio, March 7 (OSUC); 1, Davis Mts. June 2, 1937 (OSUC); 1, 7 mi. N. Sinton, May 15, 1961, (OSUC): 1, Sinton, July 31, 1964 (TAMC); 1, Corpus Christi State Park, Aug. 25, 1962 (TAMC); 1, Kingsville, July 25, 1961 (TAMC); 1, College Station, Sept. 16, 1960

(TAMC); 1, 2 mi N. Goliad, Goliad Co., June 8, 1969 (TAMC); 1, Welden Wildlife Ref., San Patricio Co. June 28, 1969 (TAMC); 1, Dallas, May 18, 1935 (TAMC); 1, Koppe's Bridge, Brazos Co., Oct. 12, 1968 (TAMC); 1, Schulenberg, June 23, 1963 (TAMC); 2, Dallas, June 9, 1906 (USNM); 6, Refugio Co., April 28, 1909 (USNM); 2, Banguete, May 15, 1942 (USNM); 1, Plano, Aug. 14, 1905 (USNM); 2, Ft. San Houston, San Antonio, (CASC). ARIZONA: Santa Cruz Co. 5, Yanks Springs, Sycamore Canyon, Tumacacori Mts., elevation 4000 feet, July 27, 1965 (CASC): 3, Procter Ranch adjacent to Madera Canyon, Sept. 20-21, 1968 (CASC): 12, Peña Blanca, Pajarito Mts. July 11-Aug. 17, 1963 (FSCA); 2, Peña Blanca, Aug. 27, 1969 (USUC); 4, Patagonia Mts., Aug. 8, 1952 (OSUC); 1, Huachuca Mts., Aug. 10, 1935 (OSUC); 1, Huachuca Mts., Aug. 19, 1950 (OSUC); 1, Patagonia, Aug. 21, 1940 (CASC); 1, Tumacacori, Sept. 1, 1931 (LACM); 4, Madera Canyon, Santa Rita Mts., Sept. 4, 1953 (LACM); 1, Madera Canyon, Santa Rita Mts., Aug. 19, 1951 (LACM); 10, Madera Canyon, Sept 28, 1963 (UDCD); 1, Sycamore Canyon, Sept 6, 1963 (UDCD); 7, Yanks Springs, 4 mi. SE Ruby, Pajaritos Mts., elevation 4000 feet, Sept 5, 1950 (AMNH); 3, 21 mi. SE. Ruby, elevation 3800 feet, Sept 5, 1950 (AMNH); 3, 61 Ranch, 15 mi. SE. Ruby, elevation 4000 feet, Sept 6, 1950 (AMNH); 5, Nogales, Aug. 4, 1947 (USNM); Pima Co. 1, Sabino Canyon, Aug. 23, 1949 (LACM); 1, 3 mi. W. Arivaca, elevation 3600 feet, Sept 4, 1950 (AMNH). Cochise Co. 1, 15 mi. NW Sierra Vista, Aug. 27, 1969 (UCDC); 1, Pearch, Aug. 26, 1961 (CISC); 2, Douglas Oct. 3-10, 1933 (CISC); 3, Douglas, Aug 2-15, 1933 (CISC); 2, Carr Canyon, Huachuca Mts. Aug. 1, 1952 (AMNH); 1, San Pedro River, 5 mi. S. Hereford, Sept. 7, 1950 (AMNH). MEXICO; DURANGO: 2, (USNM); JALISCO: 2, Lake Cahpala, July, 1940 (CASC); 2, Guadalajara, Sept. 19, 1964 (CISC); 1, Ajijic, July 5, 1962 (CISC); NAYARIT: 2, Tepic, Sept. 15, 1970 (USUC); 2, Compostela, Sept. 20, 1935 (USNM); OAXACA: 1, 4 km. E. of Tequisislan, Aug. 11, 1967, elevation 700 feet, (TAMC); SINALOA: 2, Mazatlan, Aug. 8, 1970 (USUC); 2, 14 mi. SE, Culiacan, Sept. 10, 1970 (USUC); 1, 35 mil NW Mazatlan, Sept 6, 1970 (USUC); 1, S. Lorenzo, Sept. 11, 1970 (USUC); 1, 38 Mi. E. Mazatlan, Rio Panuco, July 28, 1952 (CISC); 1, 40 mi. N. Mazatlan, July 27, 1952 (CISC). SONORA: 2, Sario, Aug. 22, 1929 (CASC); VERZCRUZ: 2, Coyame, Los Tuxtlas Range, July 1-17, 1962 (TAMC); 1, 4 mi. NE. Catelaco, Aug. 23, 1967 (TAMC); 2, Fortin de Las Flores Sumidero, elevation 3000 feet, May 17-18, 1965 (FSCA); 3, Cordoba (CASC).

Rogers (1854) described this species on the basis of a single specimen collected by Lt. H. Haldeman at Fredericksburg, Texas. Suffrian (1858) and Jacoby (1883) point out that the 4 species are very difficult to identify and their differences are obscure; they maintain the 4 as separate species. Jacoby (1883) is more inclined to treat them as a single species. *L. libatrix*, which has been synonymized, is probably of questionable status. It is believed to be more metallic green and restricted to southern Mexico and Guatemala. I have some very metallic green specimens from both southern Mexico and Guatemala and also from Texas. *L. violacescens* is a more violet species but differs little from *L. libatrix* and *L. haldemani*. Studies of the male genitalia strongly indicate that these are all 1 species. Differences in coloration are probably due to a complex of factors including host plants, moisture, and climatic conditions.

Leptinotarsa haldemani ranges from southern Oklahoma, thru Texas, and southern Arizona. The beetle has been reported in many of the states of Mexico. *L. haldemani* feeds principally on Solanaceae, especially *Physalis viscosa* and *Physalis acutifolia*. Specimen labels also record *Solanum nigrum* and *Lycopersicum esculentum* as host plants. A population of *L. haldemanı* near Austin, Texas (Bernon, 1985) feeds on *Physalis viscosa*. This population, like many of the species of *Leptinotarsa* in Texas, are active from March to May and again from August to November. Hsiao and Hsiao (1983) record *L. haldemani* feeding on *Solanum douglasii* in Peña Blanca, Arizona and in Mexico at Villa Union, Sinaloa and Tepic, Nayarit.

Leptinotarsa cacica Stal

Leptinotarsa cacica Dejean, 1836. Catalogue des Coleopteres de la collection de M. le comte Dejean. 4:421. (*Nom. Nud.*) *Leptinotarsa cacica* Stal, 1858. Ofv. Svenska Vet. Akad. Forh. 15:475

Type locality: Mexico

Jacoby, 1883:227; Tower, 1906:8; Blackwelder 1946:673.

Shinny dark blue head and pronotum, pale yellow elytra with elytra margins and scutellum of dark blue; fine, sparse punctation on head and pronotum and course punctation in irregular rows on the elytra. These features will separate this species from all others in North America. (*See back cover, center.*)

Head: shinny, dark metallic blue; punctation fine; antennae shinny, metallic blue, distal 4 segments expanded, fine pubescence; interocular distance 2.1 mm, head width 2.4 mm.

Thorax: immaculate, dark metallic blue; punctation course at lateral margins of pronotum, anterior-lateral angels of pronotum bluntly pointed; pronotum length 2.6 mm, pronotum width 6.5 mm. Scutellum metallic blue; punctation fine, dense. Elytra broadly oval, convex; punctation course, irregular, faint scratch lines running from base to apex of elytra; pale yellow with margins of elytron bluish-black, darker than head and pronotum; elytra length 11.6 mm, elytra width 9.8 mm.

Legs: unicolorous, metallic blue; femur and tibia with course, sparse punctation.

Abdomen: unicolorous, blue to a near black; punctation course, sparse.

Females.—Larger than males.

Individual variation.—Total length 15.0 mm ± 1.14 (13.0-17.5); greatest width 9.8 mm ± .86 (8.4-10.9); interocular distance 2.1 mm ± .17 (1.9-2.5); head width 2.4 mm ± .14 (2.2-2.7); pronotum length 2.6 mm ± .17 (2.4-3.0); pronotum width 6.5 mm ± .44 (5.6-7.2); elytra length 11.6 mm ± 1.0 (10.0-13.5); elytra width 9.8 mm ± .86 (8.4-10.9).

Color variation slight, elytra color varies from pale yellow to brown-yellow.

Specimens examined.—(Fig. 17, map) 27; MEXICO: 9 (USNM); DURANGO: 1, Sierra de Durango, (USNM); 6, Nombre de Dios, Aug. 13, 1947 (AMNH); OAXACA: 7, Jalapa, (AMNH); PUEBLA: 1, Huauchinango, (USNM); VERACRUX: 1, Misantla, (USNM); 2, Misantla, (AMNH).

Discussion.—This species is confined to central Mexico. Tower (1906) noted that this species was common in the savannas and foothills of Veracruz, Tabasco, and Chipas in southern Mexico. No specimens have been taken in the United States or the Mexican states bordering the United States.

L. cacica Stal has reported feeding on Compositae at Monte Blanco, near Fortin in Vera Cruz, Mexico. In August 1983 a group of entomologist including F. Drummond, R. Casagrande, R. Chauvin, T. Hsiao, J. Lashomb, P. Logan, and T. Atkinson recorded *L. cacica* parasitized by a ectoparasitic mite, *Chrysomelobia labidomerae* Eickwort. Of the 35 adult *L. cacica* collected in Vera Cruz, 86% were parasitized by his mite (1983).

Jacoby (1883) and Tower (1906) remarked about the great range in size of this species.

Leptinotarsa rubiginosa (Rogers)

Doryphora rubiginosa Rogers, 1854. Proc. Acad. Nat. Sci. Philadelphia. 8:30.

Type locality: San Antonio, Texas.

Chrysomela rubiginosa Stal, 1863. Monographie des Chrysomelides de l'Amerique. 2:168. (New generic assignment.)

Myocoryna rubiginosa Crotch, 1873. Proc. Acad. Nat. Sci. Philadelphia.

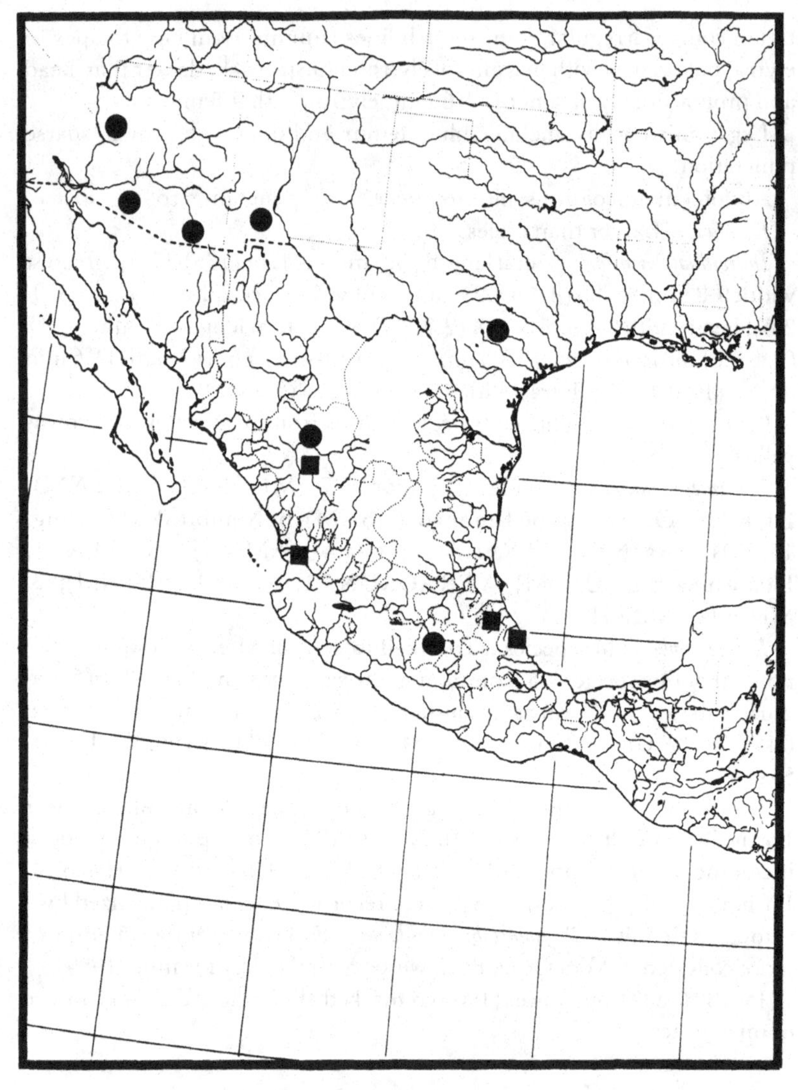

Fig. 17. Map of the distribution of *L. cacica* Stal (square), and *L. rubiginosa* (Rogers) (circle).

25:47. (New generic assignment.)

Leptinotarsa rubiginosa Jacoby, 1883. Biologia Centrali-Americana, Insecta, Coleoptera. 6(1):237. (New generic assignment.)

Suffrian, 1858:245; Linell, 1896:196; Schaeffer, 1906:230; Tower, 1906:5; Leng, 1920:294; Powell, 1941:163; Blackwelder, 1946:673, Wilcox 1972:9.

The orange-red head, pronotum, and elytra, black scutellum, antennae, and palps, orange-red ventral surface, legs either unicolorous, black or femur orange-red and remainder of leg black, are distinctive features of this species. (*See back cover, upper right.*)

Head: immaculate, orange-red; punctation fine; antennae and palps black; interocular distance 1.8 mm, head width 2.2 mm.

Thorax: immaculate, orange-red, punctation fine, coarser at lateral margins, usually sparse; anterior-lateral margins bluntly pointed; pronotum length 2.6 mm, pronotum width 5.2 mm. Scutellum black, smooth. Elytra orange-red; punctation coarse, irregular but tends to form rows; punctation darker than elytra; elytra length 7.7 mm, elytra width 7.4 mm.

Legs: unicolorous, black or with orange-red femur, sometimes part of tibia also orange-red.

Abdomen: unicolorous, orange-red.

Male genitalia.—Aedeagus cylindrical, arched; 4 times as long as wide, apex of apical area rounded (Figures 18 and 19).

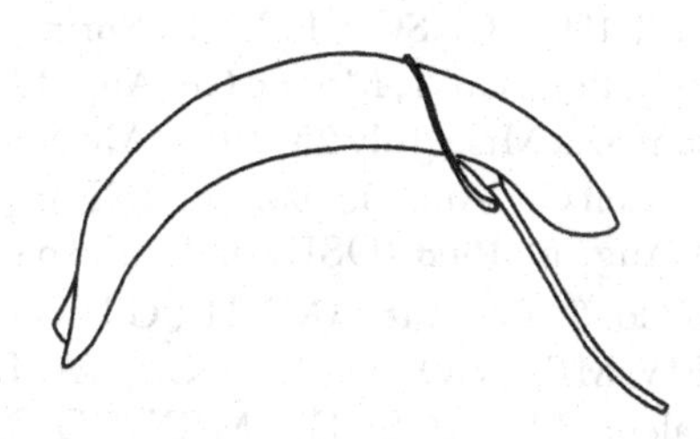

Fig. 18. Male genitalia of *L. rubiginosa* (Rogers), lateral view.

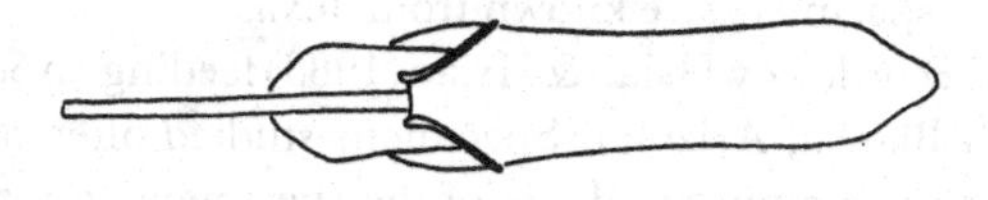

Fig. 19. Male genitalia of *L. rubiginosa* (Rogers), ventral view.

Females.—No external difference.

Individual variation.—Total length 10.5 mm ± .65 (9.3-11.9); greatest width 7.4 mm ± .43 (6.4-8.0); interocular distance 1.8 mm ± .12 (1.6-2.0); head width 2.2 mm ± .15 (1.9-2.4); pronotum 2.6 mm ± .25 (2.0-3.0); pronotum width 5.2 mm ± .28 (4.5-5.7); elytra length 7.7 mm ± .52 (6.6-8.7); elytra width 7.4 mm ± .43 (6.4-8.0).

The coloration of the head, pronotum and elytra varies from an orange-red to red, the legs seem to be the most variable, coxae are always black, femur is either black or orange-red, tibia is either black or part is orange-red and the other part is black, the underside of the body is orange-red.

Specimens examined.—(Fig. 17, map) 102: ARIZONA: 1, Madera Cn., Santa Riga Mts., July 11, 1957 (UDCD); 1, Sabino Basin, St. Catalina Mts., July 8-20, 1916 (AMNH); 1, Barnardino, Aug. 10, 1905 (AMNH); 1, Baboquivaria Mts. (OSUC); 1, Huachuca Mts. (OCUC); 2, Huachuca Mts., Aug. 1910 (CASC); Santa Cruz Co. Santa Cruz Co. 4, Peña Blanca, Aug. 29, 1969 (USUC); 3, Bear Cn. W. Nogales, Sept. 10, 1948 (USNM); 9, Ruby Rd. W. Nogales, Aug. 20, 1949 (USNM); 6, Peña Blanca, Pajarito Mts., Aug. 1, 1961 (FSCA); 44, Yanks Springs, 4 mi. SE. Ruby, Pajarito Mts., elevation 4000 feet, Sept. 5, 1950 (AMNH); 6, Nogales, Aug. 12-17, 1906 (CASC); 1, Yanks Springs, Sycamore Cn., July 28, 1965 (CASC); Pima Co. 1, Molino Cn. Aug. 12, 1958 (FSCA); 1, Browns Cn. Baboquivaria Mts., July 25, 1950 (AMNH); Cochise Co. 2, Huachuca Mts., Garden Cn. Aug. 5, 1952 (AMNH); 5, Portal, Chiricahua Mts., July 14-Aug. 6, 1968 (OSUC); Maricopa Co. 1, Phoenix, (AMNH); Yavapai Co. 2, Prescott (AMNH); Gila Co. 1, Sierra Aneho Mts. (CASC); NEW MEXICO: 1, Silver City (AMNH); 1, Cooney (OSUC); 1, Madalena Mts. (OSUC); MEXICO:DISTRITO FEDERAL: 2, San Jeronion, June 21, 1946 (AMNH); DURANGO: 3, Palos Colorados, Oct. 5, 1947 (UCDC).

Discussion.—This species has been collected in many areas of southern Arizona, some specimens studied were also from western New Mexico and Mexico, no specimens are known from Texas.

L. rubiginosa recorded by Hsiao & Hsiao (1983) feeding on *Solanum pubescens* L. in Peña Blanca, Arizona. Specimens studied often appear faded due to collecting techniques and age of the specimens in a collection. A series of 9 were recorded reared on *Physalis sp.*

Jacoby (1883) records this species as far south as the state of Puebla in Mexico.

Rogers (1854) describes this species as yellowish-brown which is probably because he had old, faded specimens. Stal (1863) describes this species as brick-red to which Jacoby objects, but this color is evident in some specimens studied. Stal also described a black discoidal spot on the thorax; Jacoby also objected to this remarking that neither he, Suffrian (1858), or Rogers (1854) came across specimens with such a spot. I have not seen any specimens with a spot on the thorax.

This species is quite distinct but there are at least two other species of Chrysomelidae in other genera that resemble *L. rubiginosa*.

Leptinotarsa defecta (Stal)

Myocoryna defecta Stal, 1859, Ofv. Svenska Vet.-Akad. Forh. 16:317.
Type locality: Texas.

Chrysomela defect (Stal, 1859) Monographie des Chrysomelides de l'Amerique. 2:165. (New generic assignment.)

Westhoff 1878:117; Jacoby 1883:234; Schaeffer 1906:239; Knab 1907:193; Leng 1920:295; Blackwelder 1946:673; Wilcox 1972:10.

The distinctive features of this species are: head and thorax pale yellow with black maculation, elytra pale yellow, sutural margin black, 2 abbreviated black vittae on each elytron, each elytron with coarse punctation in regular rows from base, to, and joining at the apex, ventral surface flavous with black maculation, legs flavous sometimes with dark areas present.

Head: pale yellow with black maculation, a black areas, sometimes U-shaped at vertex, and black areas located behind the eyes; mouth parts pale yellow; apices of mandibles black; promixal 6 segments of antennae flavous, distal 5 segments darkened and expanded. Punctation course, dense; interocular distance 1.5 mm; head width 1.8 mm.

Thorax: pale yellow with black maculation, usually U-shaped mark at the vertex and 5 spots on either side of the U mark; punctation fine at the disk, course, and dense at lateral margins; anterior-lateral angles blunt; pronotum length 2.0 mm, pronotum width 4.4mm. Scutellum pale yellow; punctation very fine. Elytra (Fig. 20) oval, convex; base color pale yellow; each elytron with 2 abbreviated black vittae, the first black vitta located near the middle of the disk, usually 3 mm long and not extending to the base or apex area; the second black vitta is located at the humerus, extending from just below the apex to the disk, usually not longer than 2 mm; sutural margin black; each elytron with 10-rows of course punctation, rows slightly irregular, first 2 join the sutural margin, others join in the apex area; elytra length 6.5 mm, elytra width 5.9 mm.

Legs: flavous, sometimes brown to black patches present, especially on

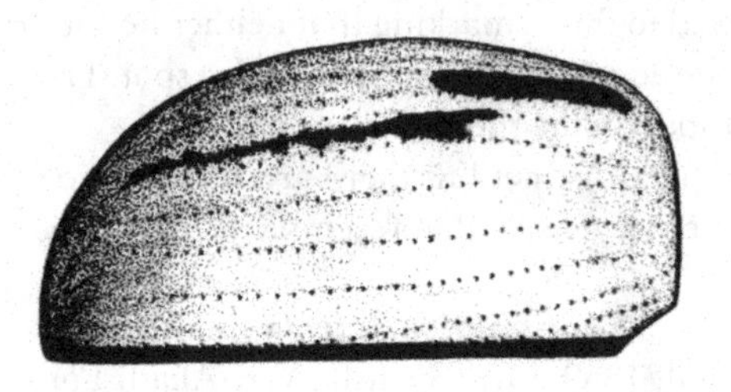

Fig. 20. Left elytron of *L. defecta* (Stal), dorsal view.

the femur.

Abdomen: pale yellow with black maculation, each sternite with a black spot at the ends and 2 black oblong patches near the center.

Male genitalia.—Aedeagus cyclindrical, slightly arched; 3.5 times as long as wide; apex of apical region rounded; base flattened, expanded (Figures 21 and 22).

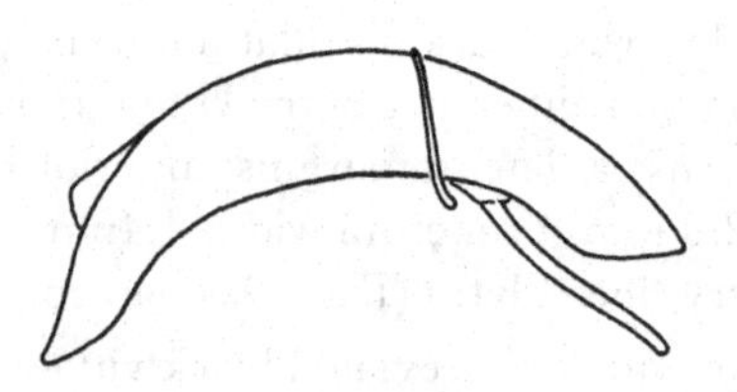

Fig. 21. Male genitalia of *L. defecta* (Stal), lateral view.

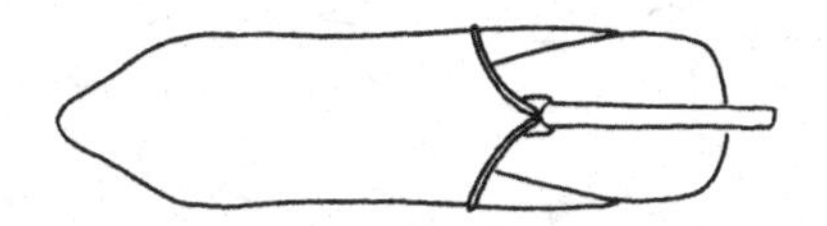

Fig. 22. Male genitalia of *L. defecta* (Stal), ventral view.

Individual variation.—Total length 8.4 mm ± .47 (7.4-9.1) greatest width 5.9 mm ± .42 (5.2-6.7); interocular distance 1.5 mm ± .08 (1.5-1.8); head width 1.8 mm ± .08 (1.7-2.0); pronotum length 2.0 mm ± .17 (1.7-2.3); pronotum width 4.4 mm ± .21 (4.0-4.7); elytra length 6.5 mm ± .41 (6.4-7.1); elytra width 5.9 mm ± .41 (5.2-6.7).

The base color of this species may be light yellow to a brownish yellow. This color variation is due in part to diet, age, and collection method. The thorax maculation is darker and more complete in some specimens.

Specimens examined.—(Fig. 23, map) 35: MEXICO: 1, "Border" April, 1939 (USNM); 1, 3 mi. S. Villagram Tamps, July 6, 1966 (TAMC); 3, Monterrey, Nuevo Leon, May 16, 1962 (TAMC); TEXAS: Cameron Co. 1, Brownsville, Sept. 8, 1943 (USNM); 4, Brownsville, May 4, 1938 (USNM); 1, San Benito, March 2, 1944 (USNM); Hidalgo Co. 3, (USNM); 1, Mercedes, March 13, 1944 (USNM); 14, Mercedes, Oct. 15-Nov. 7, 1939 (USNM); 2, Donna, Oct. 1, 1933 (TAMC); Starr Co. 4, Rio Grande City, Sept. 1, 1969 (USUC).

There has been considerable confusion between *Leptinotarsa defecta* and *L. texana*. References to *L. defecta* by Linell (1896) and Tower (1906) actually refer to *L. texana*. The two species are similar in size but the markings on the elytra are quite distinct as is the male genitalia. Both species occur in the same geographical area of southern Texas and the northern border of Mexico. *L. defecta* has been collected and observed in the three Texas border counties of Cameron, Hidalgo, and Starr.

Hsiao and Hsiao (1983) collected *L. defecta* on *Solanum elaeagnifolium* in Roma-Los Saenz, Texas. *Solanum elaegnifolium* is the major host plant of *L. defecta*. Specimen host records recorded feeding on egg-plant *Solanum melongena* and Irish potato, *Solanum tuberosum* but according to Bernon (1985) *L. defecta* "almost surely will not feed on *S. tuberosum* or *S. melongena*." Hsiao (1981) recorded no growth or reproduction for *L. defecta* on *S. tuberosum* and slight growth with no reproduction on *S. melongena*. It is safe to assume that the major host plant is the silver-leafed nightshade or *S. elaegnifolium*. Hsiao and Hsiao did report the beetle feeding on *S. tridynamum* at Merida, Yucatan, Mexico.

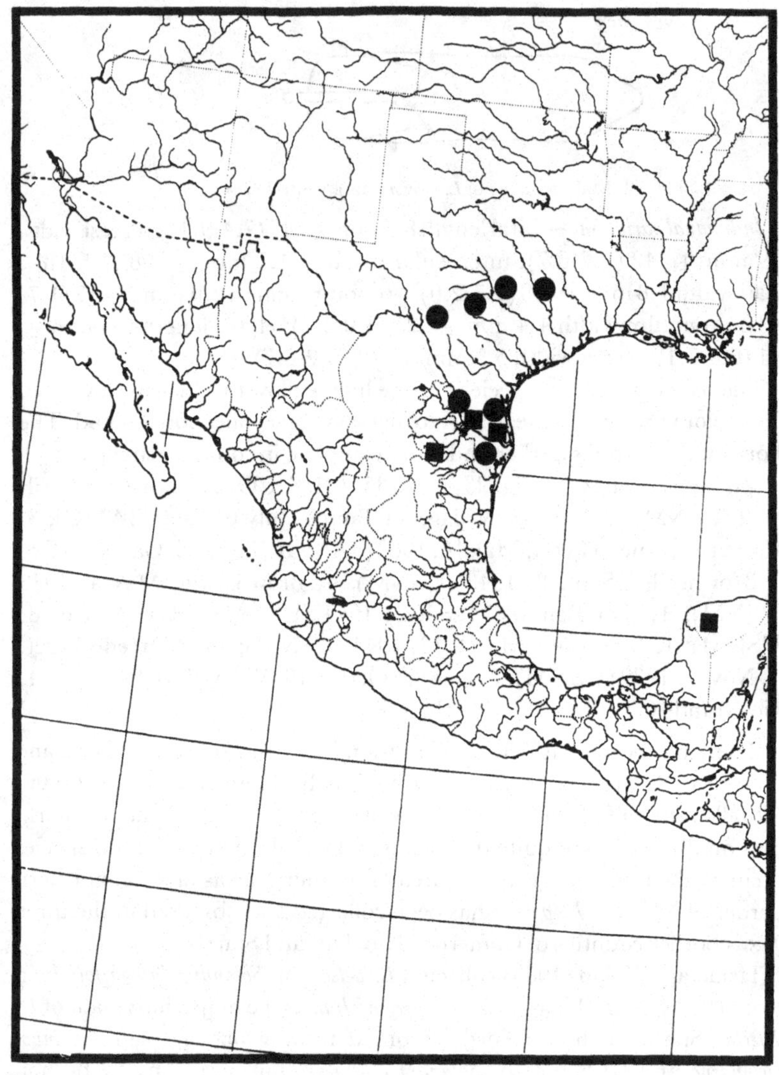

Fig. 23. Map of the distribution of *L. defecta* (Stal) (square), and *L. texana* Schaeffer (circle).

Leptinotarsa texana Schaeffer

Leptinotarsa undecemlineata texana Schaeffer, 1906. Sci. Bull. Brooklyn Inst. Arts and Sci. 1:236.

Type locality: Brownsville, Texas

Leptinotarsa texana Brown, 1961. Canadian Ent. 93:973. (New Status.)

Leng 1920:295; Blackwelder 1946:673; Wilcox 1972:10.

Head, thorax, and elytra pale yellow; head with black spot at vertex; thorax with black maculation; each elytron with four black vittae, elytra punctation course, in regular rows; ventral surface pale yellow with black maculation and flavous legs. These characters will separate this species form all others in North America.

Head: maculate, flavous-pale yellow with a black triangle at the vertex and two small black spots located behind the eyes, sometimes not visible or maybe absent altogether; mouthparts flavous, mandibles with black apices, proximal six segments of antennae flavous, distal five segments expanded and darkened; punctation variable; interocular distance 1.7 mm, head width 2.0 mm.

Thorax: maculate, pale yellow, with U-shaped mark at vertex, very variable; usually 5 black spots on either side of this U-shaped mark; punctation fine at disk, denser and courser at lateral margins; anterior-lateral angles blunt; pronotum length 2.2 mm, pronotum width 5.0 mm. Scutellum black, smooth. Elytra (Fig. 24) oval, convex; base color of elytra pale yellow, each elytron with four black vittae, each vitta extending from just below the base of the elytron to the apex, vitta 1 shorter than other 3, vitta 3 and 4 usually join at apex; punctation course, very regular, 10 rows on each elytron, first row very short and joining the sutural margin, rows 2 thru 9 border the vittae, row 10 does not border a vitta near the lateral margin; elytra length 6.4 mm; elytra width 6.2 mm.

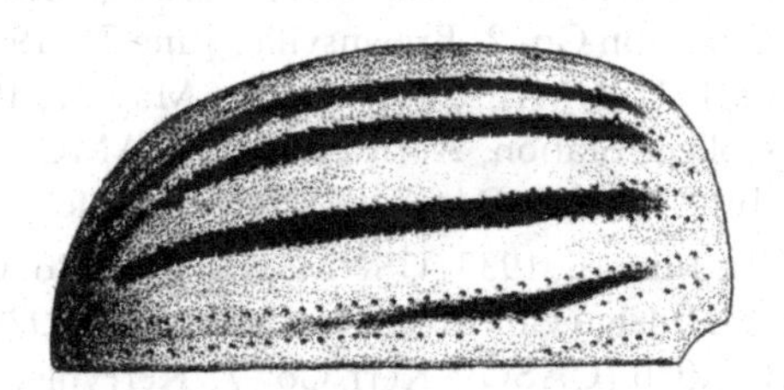

Fig. 24. Left elytron of *L. texana* Schaeffer, dorsal view.

Legs: unicolorous, flavous.

Abdomen: flavous with black maculation, sternites I-IV with 6 distinct black patches, sternite V with 2 black spots.

Male genitalia.—Aedeagus cylindrical, slightly arched; 3 times as long as wide; apex of apical region rounded; basal area flattened, expanded (Figures 25 and 26).

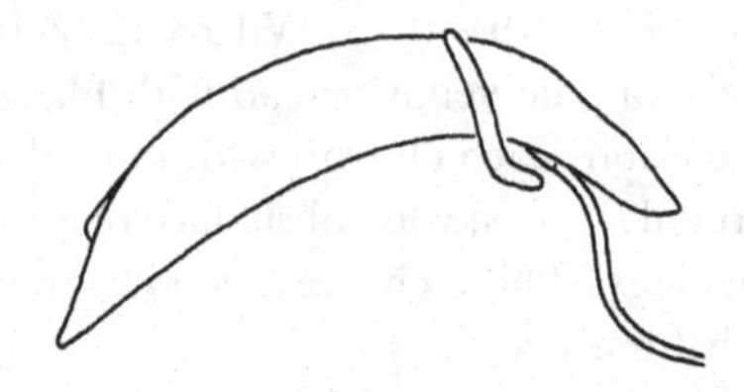

Fig. 25. Male genitalia of *L. texana* Schaeffer, lateral view.

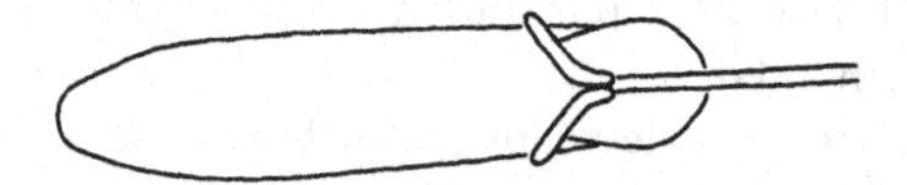

Fig. 26. Male genitalia of *L. texana* Schaeffer, ventral view.

Individual variation.—Total length 8.6 mm ± .75 (7.1-10.2); greatest width 6.2 mm ± .50 (5.3-7.0); interocular distance 1.7 mm ± .13 (1.5-1.9); head width 2.0 mm ± .13 (1.8-2.2); pronotum length 2.2 mm ± .18 (2.0-2.5); pronotum width 5.0 mm ± .36 (4.3-5.6); elytra length 6.4 mm ± .74 (5.2-7.8); elytra width 6.2 mm ± .50 (5.3-7.0).

Specimens examined.—(Fig. 23, map) 96: TEXAS, Bexar Co. 5, San Antonio, June 1942 (CASC); Brazos Co. 3, College Station, June 21, 1916 (USNM); Cameron Co. 2, Brownsville, June 25, 1930 (CASC); 4, Brownsville, June 4 (USNM); 7, Brownsville, May 21, 1910 (USNM); Brazos Co. 11, College Station, April 22, 1963 (TAMC); Comal Co. 3, New Braunfels, June 7, 1942 (CASC); 3, New Braunfels, June (CASC); 14, New Braunfels, June 16, 1933 (USNM); Eastland Co. Gorman, June 17, 1929 (CASC); Hidalgo Co. Mercedes, March 18, 1920 (USNM); 1, Mission, July 10, 1936 (CASC); Kerr Co. 7, Kerrville, Sept. 3, 1969 (USUC); Kimble Co. 2, Roosevelt, Sept 3, 1969 (USUC); Nueces Co. 18, Bishop, July 10, 1942 (USNM); Pecos Co. 1, Sheffield, Sept. 3, 1967 (USUC); Travis Co. 5, Austin, June 3, 1929 (CASC); Uvalde Co. 4, Uvalde, Sept. 1, 1969 (USUC).

According to Knab (1907) this species has generally passed among American entomologiest under the name *L. defecta* Stal, 1859. Schaeffer (1906) who described *L. texana* from Brownsville, Texas has demonstrated the distinctiveness of the two species.

Both Linell (1896), in his key to the North American *Leptinotarsa*, and

Tower (1906) have treated *L. texana* as *L. defecta*. Townsend (1903) treated *L. texana* as *L. undecemlineata*. This had led to confusion between the two species. The range of *L. texana* is larger than *L. defecta* but they do overlap in the southern tip of Texas in the area of Brownsville.

Additional confusion has occurred because *L. texana* is considered a subspecies of *L. juncta*, as indicated by Wilcox (1972). *L. texana* does resemble *L. juncta* to some extent, but *L. juncta* is a much larger species, has distinct markings, and the male genitalia are clearly different. In addition, the range is different as well as the host plants. If anything, the most often confused species are *L. juncta* with the Colorado potato beetle, *L. decemlineata*. One of the reasons the popular name "False Colorado potato beetle" has been in literature so many times is due to its confusion with the true Colorado potato beetle.

L. texana ranges in an area of southern Texas from Brownsville in Cameron county west to Pecos county and northeast on a line through Kimble county, to Austin in Travis county, and on up to College Station in Brazos county. The beetle has also been collected in Mexico along the Rio Grande valley. There is no evidence to support Neck (1983) that *L. texana* ranges from Missouri and Colorado through Texas and Mexico. *L. texana* is limited to southern Texas and the border area of Mexico. *L. defecta* has a limited U.S.A. range, being confined to an area from Cameron county west to Starr county along the Rio Grande river, but has a wider range in Mexico with records of specimens as far south as Monterrey. Hisao and Hisao (1983) have collected specimens in the Yucatan.

The main host plant for *L. texana* is *Solanum elaeagnifolium*. Hsiao (1981) reports optimal growth and reproduction on this plant, the known natural host for this species. Hsiao (1981) has also reported moderate growth and reproduction on the following Solanaceous plants: *Solanum dulcamara*, *S. carolinense*, *S. rostratum*, and *S. melongena*. It is apparent that in the range of *L. texana*, the prime host plant is the silver leafed nightshade, *S. elaeagnifolium*.

In southern Texas, according to Bernon (1987), this beetle is active from March to May and again from August to November.

Leptinotarsa tumamoca Tower

Leptinotarsa tumamoca Tower, 1918. The mechanism of evolution in *Leptinotarsa*. Carnegie Institute of Washington, no. 263:68.

Type locality: Arizona

Distinctive features: head and pronotum reddish-yellow, pronotum maculate with at least 6 black marks, elytra pale yellow, each elytron with 5 vittae, vitta 2 borders sutural margin, vitta 2, 3, and 4 join at apex of elytra, vitta 5 borders lateral margin, elytra punctation in irregular rows bordering each vittae, ventral surface, legs and abdomen reddish-yellow.

Head: immaculate, reddish-yellow; labium, maxillary palps and mandibles reddish-yellow, apices of mandibles black; distal 5 segments of antennae darkened, proximal 6 segments dark reddish-yellow; punctation course, dense near eyes; interocular distance 1.5 mm, head width 1.9 mm.

Thorax: maculate, reddish-yellow with black markings; 2 black oblong marks on either side of median line of pronotum; 2 black spots located on each lateral margin; anterior-lateral margins pointed; punctation fine, courser at the lateral margins; pronotum length 1.9 mm, pronotum width 4.2 mm. Scutellum brown, smooth, Elytra oval, convex; pale yellow, each elytron with 5 vittae; all 5 vittae originate just below the base of the elytra, vitta 1 borders the sutural margin, extends to elytra apex, vittae 2, 3 and 4 parallel along disk but join at the apex of the elytra, vitta 2 joins vitta 4 near its end and vitta 3 joins vitta 4 just above the junction of vitta 2 and 4; vitta 5 borders lateral margin; punctation course, borders each vittae, rows irregular; elytra length 6.1 mm, elytra width 5.6 mm.

Legs: unicolorous, reddish-yellow.

Abdomen: unicolorous, reddish-yellow, pubescence fine.

Male genitalia.—Aedeagus cylindrical, slightly arched; apex of apical region bluntly pointed (Figures 27 and 28).

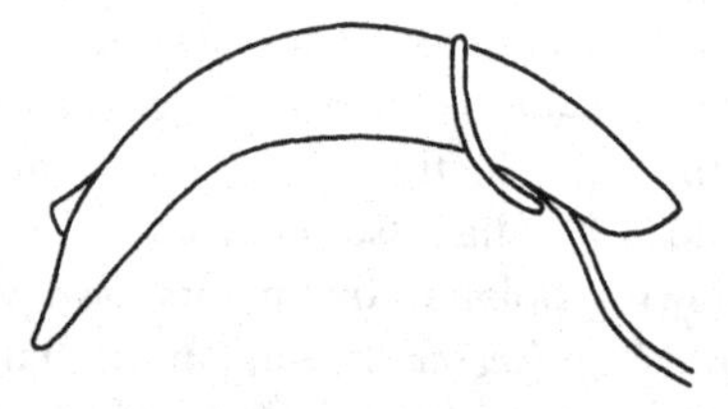

Fig. 27. Male genitalia of *L. tumamoca* Tower, lateral view.

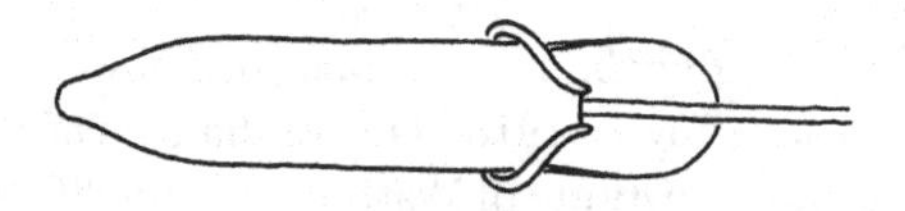

Fig. 28. Male genitalia of *L. tumamoca* Tower, ventral view.

Females.—Tower (1918) records males slightly larger than females.

Individual variation.—Total length 7.8 mm ± .33 (7.3-8.1); greatest width 5.6 mm ± .25 (5.2-5.8); interocular distance 1.5 mm ± .04 (1.5-1.6); head width 1.9 mm ± .13 (1.7-2.0); pronotum length 1.9 mm ± .13 (1.7-2.0); pronotum width 4.2 mm ± .28 (3.9-4.6); elytra length 6.1 mm ± .35 (5.6-6.5); elytra width 5.6 mm ± .25 (5.2-5.8).

Specimens examined.—(Fig. 29, map) 9: ARIZONA: Cochise Co. 1,

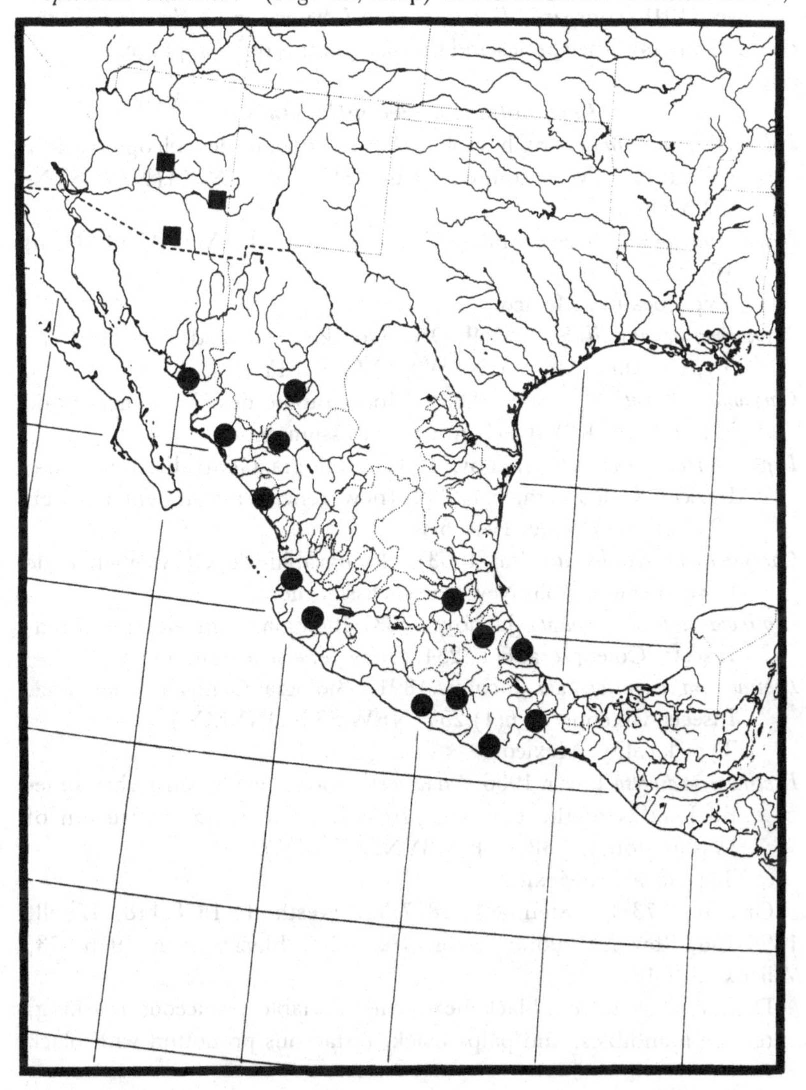

Fig. 29. Map of the distribution of *L. tumamoca* Tower (square), and *L. undecemlineata* (Stal) (circle).

Douglas, Aug. 9, 1933 (CISC); 4, 25 mi. NE. Douglas, Aug. 26, 1969 (USUC); 4, 25 mi. NE. Douglas, Sept. 1, 1970 (USUC).

L. tumamoca is confined to southern Arizona and feeds upon *Physalis acutifolia* (= *P. wrightii*) known as Wrights groundcherry. Hsiao and Hanson collected specimens in both 1969 and 1970 in the Douglas area of Arizona feeding on Wrights groundcherry. This plant is the only recorded host.

Tower (1918) compared *L. tumamoca* to *L. texana* and *L. lineolata* but the distinctiveness of the species and the host plant is quite apparent.

Leptinotarsa undecimlineata Stal

Polygramma undecimlineata Chevrolat, 1836. Catalogue des coleopteres de la collection de M. le comte Dejean. 4:421 (*Nom. Nud.*) (NEW SYNONYMY.)

Myocoryna undecemlineata Stal, 1859, Ofv. Svenska Vet.-Akad. Forh, 16:316.
Type locality: Mexico

Myocoryna signaticollis Stal, 1859. Ofv. Svenska Vet.-Akad. Forh. 16:317.
Type locality: Mexico (NEW SYNONYMY.)

Chrysomela signaticollis (Stal, 1863). Monographie des chrysomelides de l'Amerique. 4(3):163 (New generic assignment.)

Leptinotarsa signaticollis Jacoby, 1883. Biologia Centrali-Americana, Insecta, Coleoptera. 6(1):232, (new generic assignment.) Tower, 1906:6; Blackwelder 1946:673.

Chrysomela undecemlineata (Stal, 1863), Monographie des Chrysomelides de l' Amerique. 2:163 (New generic assignment.)

Leptinotarsa undecemlineata Jacoby, 1883. Biologia Centrali-Americana, Insecta, Coleoptera. 6(1):234. (New generic assignment.)

Leptinotarsa angustovittata Jacoby, 1891. Biologia Centrali-Americana, Insects, Coleoptera. 6(1):254 (NEW SYNONYMY.)
Type locality: Mexico.

Leptinotarsa diversa Tower, 1906. An investigation of evolution in chrysomelid beetles of the genus *Leptinotarsa*. Carnegie Institution of Washington, no. 48. (NEW SYNONYMY.)
Type locality: Mexico.

Crotch 1873:47; Steinheil, 1877:32; Westhoff, 1878:118; Linell, 1896:196; Tower, 1906:5; Leng, 1920:295; Blackwelder, 1946:673, Wilcox 1972:10.

Distinctive features: black head with variable testaceous marking, antennae mandibles, and palps black, testaceous pronotum with black maculation, black scutellum, elytra testaceous, each elytron with four distinct black vittae, punctation course, irregular rows underside completely black with fine white pubescence. (*See back cover, lower left.*)

Head: maculate, black with testaceous maculation, variable; testaceous area in a C-shaped area, bottom of the C just above clypeus and top near vertex; open part of C faces mid-line of head, testaceous area greatly expanded giving the appearance of a testaceous head rather than black; area between C's is black and can take the form of a triangle, circle or spot; punctation variable; antennae, palps, and mandibles black; interocular distance 1.8 mm; head with 2.0 mm.

Thorax: maculate, testaceous with a large black U-shaped or horseshoe mark at center of pronotum, small black dot located near anterior-lateral angle, large irregular black area located on either side of horseshoe near occiput; this area can be split up into smaller blotches or spots; anterior-lateral angles blunt; punctation variable, coarser near lateral margins; pronotum length 2.2 mm, width 5.0 mm. Scutellum black, smooth. Elytra (Fig. 30) pale yellow, outlined in black; each elytron with 5 vittae; vitta 1 shorter than other 4 and adjacent to the sutural margin; vittae 2 thru 5 extend more than half the length of the elytron and are very distinct; vitta 2 joins the sutural margin 34 the way down the elytron, vitta 3 and 4 usually join at apex, apical margin thickened, resembling a sixth vitta; punctation coarse, in irregular rows bordering each vittae and lateral margins; elytra length 8.2 mm, elytra width 6.9 mm.

Fig. 30. Left elvtron of *L. undecemlineata* (Stal), dorsal view.

Legs: unicolorous, black; pubescence fine, white.

Abdomen: unicolorous, black; pubescence fine, white.

Male genitalia.—Aedeagus cylindrical, nearly straight; 4 to 5 times as long as wide; apex broadly rounded; basal area slightly flattened (Figures 31 and 32).

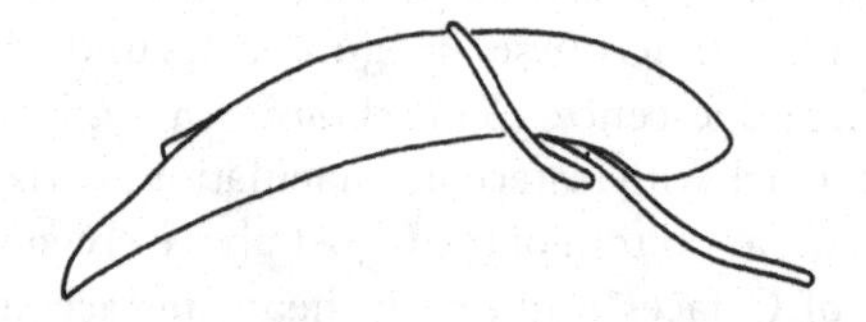

Fig. 31. Male genitalia of *L. undecemlineata* (Stal), lateral view.

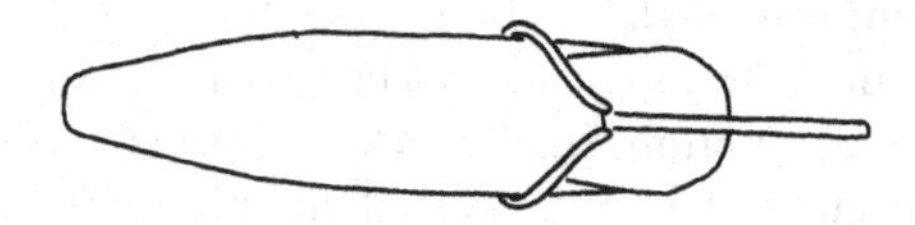

Fig. 32. Male genitalia of *L. undecemlineata* (Stal), ventral view.

Females.—gravid females much larger than males.

Individual variation.—Total length 10.8 mm ± 1.4 (8.9-13.1); greatest width 6.9 mm ± .82 (5.6-8.4); interocular distance 1.8 mm ± .16 (1.5-2.0); head width 2.0 mm ± .19 (1.8-2.5); pronotum length 2.2 mm ± .19 (2.0-2.5); pronotum width 5.0 mm ± .40 (4.4-5.9); elytra length 8.2 mm ± .95 (7.0-10.0); elytra width 6.9 mm ± .83 (5.6-8.4).

Specimens examined.—(Fig. 29, map) 333: TEXAS: 1, San Antonio, May 24, 1941 (CASC); MEXICO: CHIAPAS: 1, 31 km. E. La Trinitaria, Aug. 14, 1967 (TAMC); 1, 19 km. E. Teopisca, Aug. 15, 1967 (TAMC); 1, 1.5 mi. E. Campete, March 7, 1953 (CISC); 1, Simojove, March 16, 1953 (CISC); 1, 15 mi. NW. Comitan, Aug. 3, 1952 (CISC); CHIHUAHUA: 1, Bacuchiac, Aug. 27, 1950 (OSUC); 12, Bacuchiac, Aug. 27, 1950 (CISC); JALISCO: 3, Jamay, Aug. 24, 1966 (TAMC); 1, Sebastian, Sierra Madeiro Mts. (CASC); 15, Ajijie, July 5, 1962 (CISC); 3, Puerto Vallarta, May 30, 1967 (CISC); 4, 6 mi. W. Chapala, June 30, 1963 (CISC); 3, Chapala, July 5, 1962 (CISC); 1, Zapotlaneojo, June 18, 1903 (USNM); NAYARIT: 1, 25 mi. SE tepic, Sept. 15, 1970 (USUC): 2, 23 mi. SE Tepic Aug. 21, 1964 (TAMC) 1, 10-15 mi NE of San Blas on Rt 54, elevation 0-500 ft., Aug. 29, 1963 (FSCA): 5 mi E. San Blas, Sept. 3, 1957 (ISUC): 1, San Blas, Sept. 17-21, 1953 (CASC): 66, vicinity of Compostela, June 20-July 10, 1933-1936 (LACM); 2, Puerta de la Lima, Sept. 10, 1950 (LACM); 1, 8.6 mi. N. Compostela, Oct. 5, 1950 (CISC); OAXACA: 4, Monte Alban Ruins, 8 mi. SW. Oaxaca, elevation 6000-6500 feet (FSCA): 5, Juquila Mixes, 4700 feet, July, 1958 (HAHC); 4, Oaxaca, June 15, 1968 (HAHC); 4, 12 mi. SE. Nochixtlan, Dec. 13, 1948

(CASC); 1, Monte Alban, Dec. 11, 1948 (CASC); 1, 12 mi. SE. Nochixtlan, Dec. 13, 1948 (CASC); 7, 64 mi. W. Tehuantepec, July 21, 1952 (CISC); 1, Monte Alban Ruins, Aug. 3, 1964 (CISC); 1, 20 mi. E. El Cameron, July 21, 1956 (CISC); 1, 9.5 mi. E. Rio Amapa, Temascal (CISC); PUEBLA: 2, Villa Juarez, March 24, 1954 (CISC): SAN LUIS POTOSI: 1, 1 mile SW. Tamazunchale, July 7, 1966 (TAMC); 5, Tamazunchale, July 10, 1937 (INHS): SINALOA: 9, 50 mi. NE. Mazatlan, Sept. 9, 1970 (USUC); 2, 38 mi. E. Jct. Hwy. 15 & 40 on Hwy. 40, 11000 feet, Aug. 26, 1964 (CASC): SONORA: 4, 25 mi. S. Navojoa, Sept. 13, 1963 (CISC): TABASCO: 1, 14 mi. SE. Villahermosa, Sept. 8, 1968 (TAMC): VERACRUZ: 18, Lake Catemaco, Catemaco, June 27, 1965 (OSUC): 5, Lake Catemaco, Coyame, Aug. 8, 1963 (FSCA); 7, 5 mi. N. Lerdo de Tejade, Aug. 23, 1967 (TAMC); 1, Sontecomapan, July 7, 1962 (TAMC): 3, San Andres, Tuxtla, Aug. 5, 1962 (TAMC); 1, Ebsat Los Tuxtlos Range, Aug. 2, 1963 (TAMC); 1, 5 mi. S. Nautla, June 29, 1965 (TAMC); 2, 4 mi. NW. Sontecomapan, June 9, 1965 (TAMC); 22, 7 mi N. Lake Catemaco, Hotel Playa Azul, Aug. 16, 1963 (FSCA); 2, Lake Catemoca, Hotel Playa Azul, July 30, 1965 (FSCA); 23, 10 mi. E. of Catemaco, Dec. 4, 1955 (ISUS); 4, Orizaba, Sept. 22, 1897 (CASC); 7 mi. N. Cantiago, Tuxtla, July 8, 1963 (CISC): CANAL ZONE: 14, from Corn Fodder on Barge, Limon to Juan Mina Aug. 31, 1918 (USNM). 3, Corozel, Oct. 26, 1918 (USNM); 2, Gamboa, Aug. 31, 1918 (USNM); 3, Mindi, July 10, 1918 (USNM); COSTA RICA: 1, Turrialba Aug. 7-9, 1965 (LACM) EL SALVADOR: 7, Cerro Verde, 6800 ft., July 19, 1963 (CASC); GUATEMALA: 1, Antigua, Sept. 1949 (ISUC); 2, Panajachel, Aug. 19, 1968 (CASC); 1, Santa Itena, July 31, 1926 (CASC); 3, Santa Rita, 10 mi. E. Nahuala Solola, Sept. 3, 1965 (CISC); HONDURAS: 1, San Pedro, Feb. 21, 1908 (OSUC); CUBA: 5, Gran Piedra nr. Santiago, Oriente Prov., May 30-31, 1959 (INHS); 1, Loma (Pico) del Gato, Sierra Maestra, Oriente Prov., May 26-28, 1959 (INHS); 6, Vinales, PInar del Rio, July 17, 1948 (UCDC); SOUTH AMERICA: COLUMBIA: 8, Bogota (LACM).

Jacoby (1883) cites many localities in Mexico, Central, and South America for this species. Tower (1906) discusses its wide distribution second only to *L. decemlineata*. The range of this species is from Chihauhua and Sonora in northern Mexico, east to Cuba and south to Colombia, South America. This species is not found in the United States. There have been some reports of *L. undecemlineata* in the United States but no conformation. There are a number of host plants for this species, all in the family Solanaceae. One series of adults were labeled as feeding on *Solanum mitlese*. Hsiao and Hsiao (1983) collected pupae on *Solanum lanceolatum* Cav. at Jalapa, Veracruz, Mexico. They also collected pupae on *Solanum ochraceoferrugineum* (Dunal) Fern. at Lake Catemaco, Veracruz, and Lake Chapala, Jalisco, both in Mexico.

Logan (1983) collected adults and larvae from August 21 to September 12 on *S. ochraceoferrugineum* in a number of localities in Mexico including: 10 km south of Cuernavaca, Morelos; 8 km east of Tuxpan; 21 km east of Morelia; 29 km east of Uruapan; 7 km south of Uruapan, and 3 km south of Samora, all in the state of Michoacan. In addition, adults were collected in Chapala; southeast of Puerto Vallarta in Jalisco; Oaxaca in Oaxaca; 55 km east of Tuxtla Gutierrez, and 7 km west of San Cristobal in Chiapas. Adults were also collected at Aqua Blanca Rd., in Tabasco; Lake Catemaco, San Andres Tuxtla, Lake Catemco, Jalapa, La Estanzuda, and Tuzamapan, all in Veracruz.

Hsiao (1983) also recorded numerous larvae parasitized by a tachinid, *Doryphorophaga* sp. (Diptera: Tachinidae). Drummond *et. al* (1984) reported a mite, *Chrysomelobia labidomerae* Eickwort (Acari: Tarsonemina; Podapolipidae), under the elytra of three species of *Leptinotarsa* including *L. undecnlineata*. In one collection of 20 adults collected at Morelia in Michoacan, 75% of the adults were parasitized. Parasitization ranges from 0% to 100% throughout the collecting sites visited by Logan *et. al.* in 1983.

It should be noted that *Leptinotarsa signaticollis* (Stal) is a synonym of *L. undecemlineata*. After considerable examination of specimens and observations made in the field by Logan (1983) it has become clear that this species differs only in markings on the elytra. These markings are identical to *L. undecemlineata* except that they are greatly reduced in number giving an appearance of numerous black spots. If one begins to connect the ''dots'' as in the ''connect a dot game'' one comes up with *L. undecemlineata*.

Laptinotarsa juncta (Germar)

Chrysomela juncta Germar, 1824. Insectorum species novae aut minus cognitae. p. 590.
Type locality: Georgia.

Polygramma juncta Dejean, 1836. Catalogue des Coleopteres de la collection de M. le comte Dejean. 4:421.
(New generic assignment.)

Doryphora juncta Rogers, 1854. Proc. Acad. Nat. Sci. Philadelphia. 8:30.
(New generic assignment.)

Chrysomela juncta Stal, 1963. Monographie des Chrysomelides de l'Amerique. 2:165. (New generic assignment.)

Myocoryna juncta Crotch, 1873. Proc. Acad. Nat. Sci. Philadelphia. 25:47. (New generic assignment.)

Leptinotarsa juncta Linell, 1896. Journ. New York Ent. Soc. 4:196. (New generic assignment.)

Suffrain, 1858:243; Motschulsky, 1860:181; Westhoff, 1878:118; Tower, 1906:7; Blatchley, 1910:1154; Leng, 1920:294; Wilcox, 1954:413; Wilcox, 1972:10; Arnett, 1980:326; Arnett 1985:376.

Distinctive features: rufescent head and pronotum with black maculation; scutellum darkened; elytra flavous, each elytron with 5 black vittae, vittae 3 and 4 join at apex, space between vittae 3 and 4 darkened, elytra punctation in very regular rows; femur with black spot.

Head: maculate, rufescent; two small black spots on the clypeus, sometimes lacking; one large irregular mark, sometimes triangular, on the vertex, always present; two small spots just below the larger mark and off center but sometimes lacking; antennae and palps rufescent; apices of mandibles darkened; punctation variable; interocular distance 2.0 mm, head width 2.4 mm.

Thorax: maculate, rufescent; anterior-lateral angles blunt; 2 large, irregular, elongated black spots on either side of median line of pronotum, sometimes connected at the base and forming a horseshoe; small black spot at the base of horseshoe just above basal margin of elytra; usually 6 spots located on either side of median line of pronotum, sometimes a few are confluent; punctation sparse, fine, coarser and denser at lateral margins; pronotum length 2.6 mm, pronotum width 5.4 mm. Scutellum darkened, smooth. Elytra (Fig. 33) oval, convex, flavous; elytron with 5 black vittae;

Fig. 33. Left elytron of *L. juncta* (Germar), dorsal view.

vitta 1, bordering sutural margin, extends from just below the base to apex; vitta 2 shorter than first, does not reach base; vittae 3 and 4 connect at apex of elytron; space between black; vitta 5 along lateral margin of elytron; punctation course, in very regular rows outlining each vittae; 2 short rows of punctures on either side of scutellum and extending 1/4 way down elytron; elytra length 7.9 mm elytra width 7.3 mm.

Legs: rufescent; punctation coarse, sparse; pubescence fine, golden; black spot located on outer margin of femur.

Abdomen: rufescent, with 6 black discoidal spots on sterna I-V and 2 black spots on sterna VI.

Male genitalia.—Aedeagus cylindrical, gently arched; 4 times as long as wide; apex a flattened point (Figures 34 and 35).

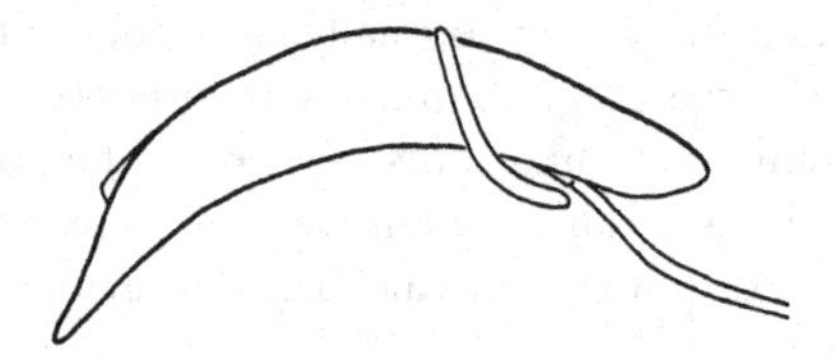

Fig. 34. Male genitalia of *L. juncta* (Germar), lateral view.

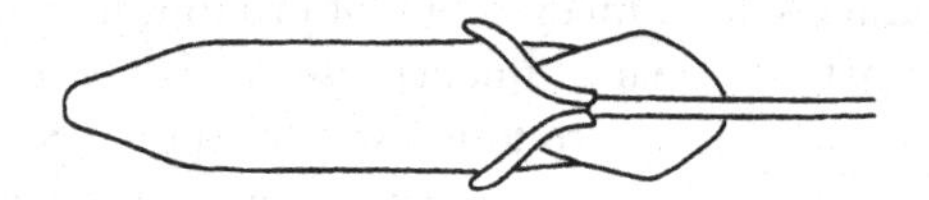

Fig. 35. Male genitalia of *L. juncta* (Germar), ventral view.

Females.—Gravid females larger than males.

Individual variation.—Total length 10.6 mm ± .92 (8.3-12.0); greatest width 7.3 mm ± .67 (5.3-8.0); interocular distance 2.0 mm ± .15 (1.5-2.2); head width 2.4 mm ± .15 (1.9-2.6); pronotum length 2.6 mm ± .23 (2.3-3.0); pronotum width 5.4 mm ± .34 (4.5-5.8); elytra length 7.0 mm ± .70 (6.3-8.6); elytra width 7.3 mm ± .67 (5.3-8.0).

The head and pronotum maculation vary both in size of spots and lack of them. The area between vittae 3 and 4 is not always totally darkened but usually more than 75% blackened. Base color of elytra sometimes more rufescent than flavous.

Specimens examined.—(Fig. 36, map) 71: ALABAMA: 2, (AMNH): 1, Auburn, April 19, 1940 (CASC): FLORIDA: 3, (AMNH); 2, Winter Park, April 16, 1938 (AMNH); GEORGIA: 1, Atlanta, July 17, 1953 (CASC); 1, Atlanta, July 28, 1940 (UDCD); 3, Myrtle, Aug. 3, 1906 (USNM); 4. Ft. Valley, April, 1903 (USNM); 1, East Point, 1933

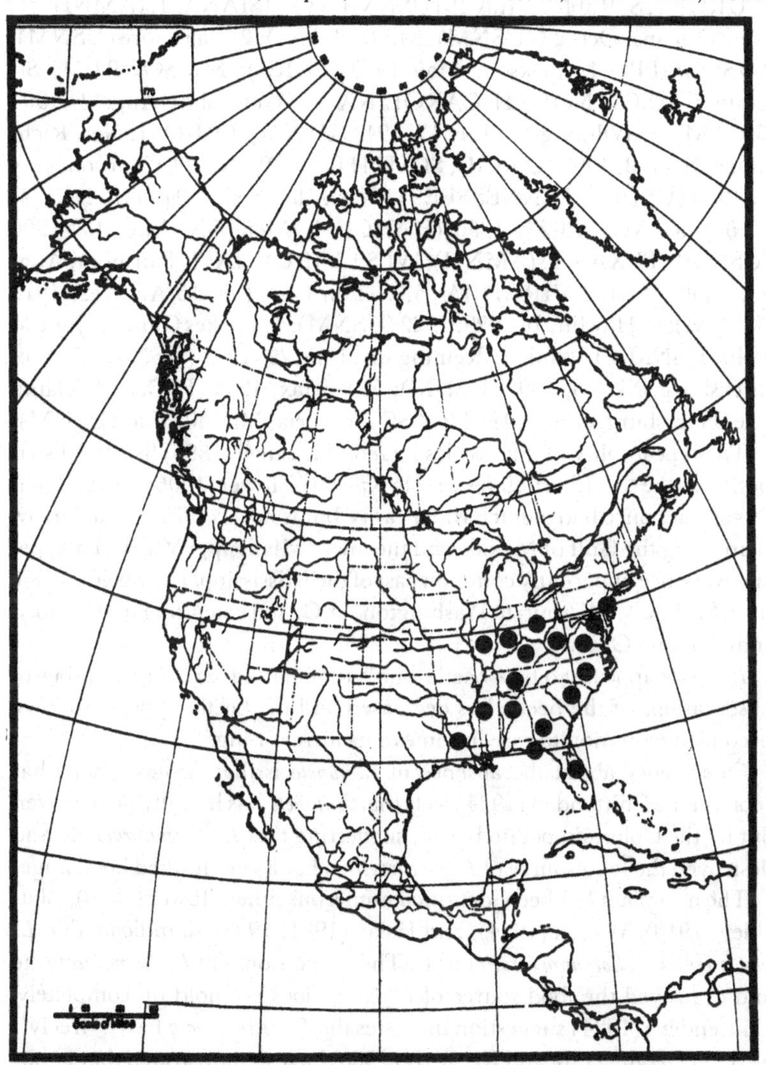

Fig. 36. Map of the distribution of *L. juncta* (Germar) (circle).

(USNM); ILLINOIS: 1, McLeansboro, June 11, 1910 (USNM): INDIANA: 1, Lawrence Co., July 20, 1908 (USNM); 4, Wetunka, (USNM); 1, Tippecanoe Co., July 5, 1931 (PURC); 1, Posey Co., April 21, 1909 (PURC); 1, Knox Co., June 10, 1904 (PURC); 1, Putnam Co., July 23, 1909 (PURC); KENTUCKY: 1, Campbell Co., Aug. 4, 1931 (PURC); 1, S. Robling, July 19 (USNM); LOUISIANA: 1, (AMNH); 1, New Orleans, Oct. 23 (USNM); MARYLAND: 2, Baltimore (USNM); MISSISSIPPI: 1, Jackson, May 1950 (PURC); MISSOURI, 1, St. Louis (CASC); NORTH CAROLINA: 1, Westpoint, Aug. 2, 1904 (USNM); 1, Wilmington, July 13, 1942 (USNM); OHIO: 1, New Richmond, Sept. 8, 1907 (USNM); SOUTH CAROLINA: 1, Clemson, July 4, 1930 (USUC); TENNESSEE: 1, Nashville, Sept. 1943 (UCDC); 4, Knoxville, May 18-23, 1956 (HAHC); 1, Memphis, May 15, 1906 (USNM); TEXAS: 2, (AMNH); WEST VIRGINIA: 1, Fairmont, Aug. 23, 1930 (PLID); VIRGINIA: 5, Nelson Co., July 15-Aug. 12, 1916 (USNM); 4, Hawlin, June 29, 1922 (USNM); 1, Burkes Garden, June 5, 1940 (USNM); 1, 4 mi. S. Remington, June 29, 1947 (USNM); 5, Fredricksburg, May 8, 1892 (USNM); 5, Fairfax, 1953 (FSCA); 1, Camp Lee, Petersburg, June 1, 1944 (CASC); 1, Syria, May 30, 1959 (USNM).

The type locality of this species is Georgia. Linell (1896) lists the distribution as the "southern states west to Kansas." Tower (1906) remarks that this species once had a much wider range but is now confined to a narrow strip along the Gulf of Mexico and the lower Mississippi Valley. This species was once apparently common east of the Mississippi river below a line from St. Louis, Missouri to Washington, D.C. and extending into Illinois, Indiana, and Ohio.

L. juncta appears to be a beetle who's time has passed. The number of observations of the beetle has decreased and the beetle appears to have become uncommon to rare in some of its former range.

One theory about the absence of *L. juncta* is that *L. decemlineata* has replaced it. Chittenden (1924) feels that *L. juncta* has literally been driven out by the Colorado potato beetle, suggesting that *L. decemlineata* ate and destroyed the food source of *L. juncta* which has a very limited host range.

The host plant has been identified numerous times, Tower (1906), Blatchley (1910), Wilcox (1954), and Hsiao (1981, 1983) all indicate that *L. juncta* feeds on *Solanum carolinense* L. The suggestion that *L. decemlineata* ate and destroyed the food source of *L. juncta* does not hold up completely. Chittendens (1924) suggestion indicates the *L. decemlineata* feeds "freely" on *S. carolinense*. Hare and Kennedy (1986) have demonstrated that populations of *L. decemlineata* differ widely in their ability to survive on *S. caro-*

linense. Populations from North Carolina survived best while populations from Virginia, New Jersey and Connecticut had very low survival. There appears to be little host plant competition between the two beetles. Neck (1983) reviews Towers (1906) theory that there was interbreeding between *L. decemlineata* and *L. juncta* and this resulted in genetic swamping and near extinction of *L. juncta*.

Whatever the reasons for *L. juncta* to become less common in its former range, it is, however, found on its host plant in the southeastern states. Bailey and Kok (1978) reported *L. juncta* as being associated with *S. carolinense* throughout Virginia.

It may not be as common as previously reported but an additional factor needs to be at least mentioned and that is misidentification. *L. juncta* and *L. decemlineata* have probably been confused by many workers in the field. The quickest way to tell these two species apart is by checking the punctation on the elytra, which in *L. juncta* is very regular and very irregular in *L. decemlineata*. In addition, the presence of a black spot on the outer margin of the femur is also a good spot identifying character of *L. juncta*.

Another unfortunate aspect of the *L. juncta* story is the choice of a common name. In many earlier publications the name "False Colorado potato beetle" or "False potato beetle" have been used. The Entomological Society of America has approved "False Potato Beetle" as the official common name for *L. juncta* (1978). This is unfortunate since the general public may get the idea that this beetle is often present on cultivated potatoes. Both Arnett (1985) and Jacques (1985) use the name Horse-nettle beetle in publications. This would be a preferred name since it clearly indicates the host plant on which the beetle is most likely feeding. One reason for less sightings of the beetle may very well be that not many people collect on the spiny horse-nettle plant, while every gardener that sees a potato beetle reports it to local extension agents.

Leptinotarsa decemlineata (Say)

Doryphora decemlineata Say, 1824. Journ. Acad. Nat. Sci. Philadelphia 3(3); 453.
Type locality: Missouri and Arkansas. *Myocoryna multilineata* Stal, 1859. Ofv. Svenska Vet.-Akad. Forh. 16:316. (Synonymy, Stal 1863:164.)
Type locality: Mexico

Myocoryna multitaeniata Stal, 1859. Ofv. Svenska Vet.-Akad. Ford. 16:317. (NEW SYNONYMY.)
Type locality: Mexico

Chrysomela decemlineata Stal, 1865. Monographie des Chrysomelides de l'Amerique. 3:322. (New generic assignment.)

Leptinotarsa decemlineata Kraatz, 1874. Berliner Ent. Zeitschr. 18:442. (New generic assignment.)

Leptinotarsa intermedia Tower, 1906. An Investigation of Evolution in chrysomelid beetles of the genus *Leptinotarsa*. Carnegie Institution of Washington. 48:6. (NEW SYNONYMY.)
Type locality: Mexico

Leptinotarsa oblongata Tower, 1906. An Investigation of Evolution in chrysomelid beetles of the genus *Leptinotarsa*. Carnegie Institution of Washington. 48:6 (NEW SYNONYMY.)
Type locality: Mexico.

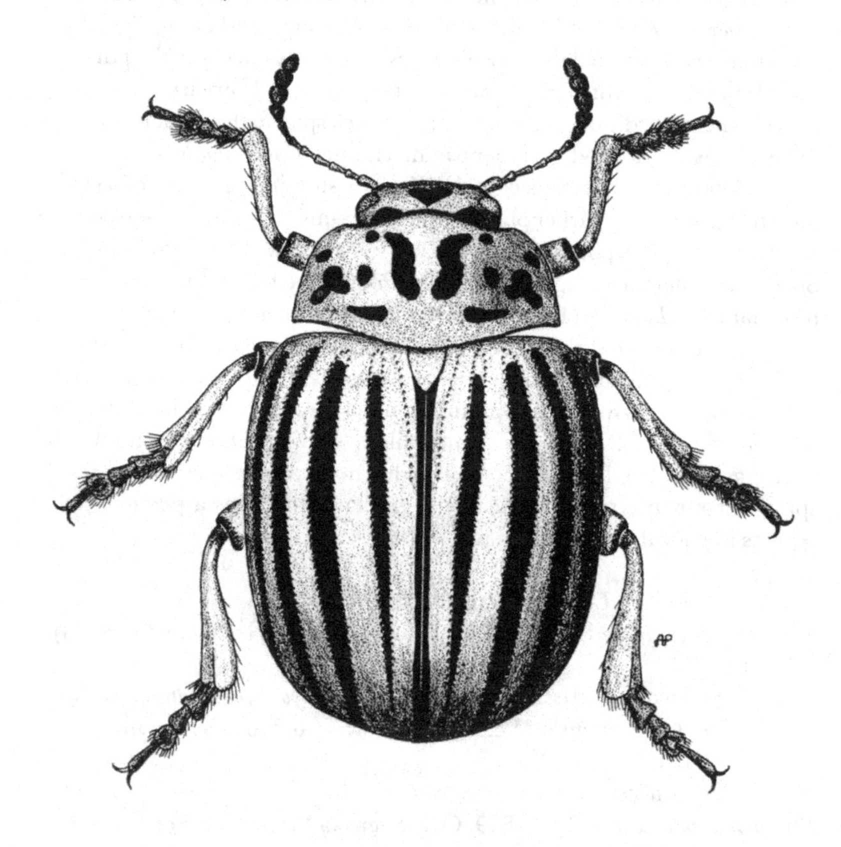

Fig. 37. Adult of *L. decemlineata* (Say), dorsal view.

Leptinotarsa rubicunda Tower, 1906. An Investigation of Evolution in chrysomelid beetles of the genus *Leptinotarsa*. Carnegie Institution of Washington. 48:7. (NEW SYNONYMY.)
Type locality: Mexico.

Rogers, 1854:30; Suffrian, 1858:244; Shimer, 1866:84; Crotch, 1873:47; Kraatz, 1878:131; Westhoff, 1878:113; Jacoby 1883:233, Townsend, 1886:57; Jacoby, 1891:253; Linell, 1896:196; Tower, 1906:7; Knab, 1907:190; Blatchley, 1910:1154; Leng, 1920:294; Metcalf and Flint, 1939:507; Powell, 1941:163; Wilcox, 1954:413; Borrow and White, 1970:200; Wilcox 1972:10; Arnett 1980:326; Arnett 1985:376.

Head, thorax, and elytra pale yellow to flavous, head and thorax with black maculation; thoracic epipleura flavous; each elytron with 5 black vittae, elytra punctation course, in irregular rows, ventral surface pale yellow to flavous with black maculation; legs pale yellow to flavous, joint areas and tarsal segments darkened to black. This will separate the species from all others in North America.

Head: maculate, pale yellow to flavous with black mark, usually triangular at the vertex and 2 black spots located behind the eyes usually concealed by the pronotum; mouthparts pale yellow to flavous; mandibles with apices black; terminal segment of maxillary palpus darkened; antennae with proximal 6 segments flavous, distal 5 segments expanded and darkened; punctation variable; interocular distance 1.7 mm, head width 2.0 mm.

Thorax: maculate, pale yellow to flavous; U-shape mark or 2 longitudinal black marks at the vertex with 5 black spots on either side, sometimes confluent; punctation fine at disk, coarser and denser at lateral margins; anterior-lateral angles rounded to blunt; epipleura flavous, sometimes darkened; pronotum length 2.2 mm, pronotum width 5.0 mm. Scutellum yellow to black, smooth.

Elytra oval, convex; base color of elytra pale yellow to flavous, each elytron with 5 black vittae, all 5 vittae extend from base to apex of elytra; sutural margin black; vitta 1 nearly joins sutural margin at apex, vittae 2 and 3 join at apex, vittae 4 closer to 3 than 5, vittae 5 adjacent to lateral margin of elytra; punctation course, in irregular rows along vittae; elytra length 7.8 mm, elytra width 6.8 mm.

Legs: pale yellow to flavous; femur-tibia joint darkened, tarsal segments brown to black.

Abdomen: pale yellow to flavous; sterna I-V with black spot at lateral margins; sterna I-IV with black oblong spots on either side of abdomen midline.

Male genitalia.—Aedeagus cylindrical, greatly arched; apex flattened; aedeagus 3-1/2 times as long as wide (Figures 9 and 10).

Females.—Gravid females considerably more robust than males and non-gravid females.

Individual variation.—Total length 10.0 mm ± .62 (9.0-11.5); greatest width 6.8 mm ± .42 (6.1-7.6); interocular distance 1.7 mm ± .13 (1.5-2.0); head width 2.0 mm ± .13 (1.7-2.3); pronotum length 2.2 mm ± .23 (1.7-2.6); pronotum width 5.0 mm ± .25 (4.7-5.7); elytra length 7.8 mm ± .50 (7.0-8.7); elytra width 6.8 mm ± .42 (6.1-7.6).

Base color of *L. decemlineata* can vary from pale yellow to a brown yellow. Thorax maculation also varies from a very distinct number of spots to confluent spots.

Leptinotarsa decemlineata is the most widespread species of the genus. It is commonly called the Colorado potato beetle. This species occurs not only in North America but can now be found in most parts of Europe and as far east as portions of the Soviet Union and China.

There is little doubt among entomologists that the Colorado potato beetle has to rank near the top of a list of most destructive insect pests for the world.

The life cycle of the beetle is well known. Adults over winter buried in the soil; in the spring; the beetles emerge from the soil and begin to feed on available host plants. The female lays up to 500 eggs in clusters each averaging about 20 eggs. Egg laying lasts from 1 week to a month. Adults die within 5 weeks after they emerge from the soil. Egg hatching occurs about 1 week after oviposition. The small, cryphosomatic, reddish larvae attack the leaves of host plants. The larvae pass through 4 instars and become full grown in 15 to 20 days. The larvae then drop from the plants and burrow into the soil where they construct a spherical cell and transform into a yellowish pupal stage which lasts from 5-10 days. The adult emerges and after feeding for a few days lays eggs for the second generation. Two generations per year are common, although there may be only a single generation in northern United States and Canada and sometimes a partial third generation in the southern states. The beetle has been recorded from northern and central Florida but has not been detected in the southern counties south of Lake Okeechobee.

There are a number of host plants reported for the Colorado potato beetle in addition to its original host plant *Solanum rostratum* and the cultivated potato, *S. tuberosum*. Barber (1933) reports *Solanum sisymbriifolium* being attacked in eastern Georgia. Brues (1940), in studies on food preferences of the beetle, found that the potato was preferred but other species of Solanaceae were attacked including: *Solanum melongena*, (egg-plant), *S.*

dulcamara, *S. subinerme*, *S. marginatum*, *Lycopersicum esculentum* (tomatoes), *Physalis* spp. (groundcherry), and *Nicotiana tabacum* L. (cultivated tobacco). Hisao (1981) recorded optimal growth and reproduction for *L. decemlineata* on the following known natural host: *S. tuberosum*, *S. dulcamara*, *S. rostratum*, and *S. melongena*. He recorded moderate growth and reproduction of the following natural hosts: *S. carolinense*, *S. elaeagnifolium*, *S. sarachoides*, *Hyoscyamus niger*, and certain varieties of *Lycopersicon esculentum*.

In addition to the extensive range of *L. decemlineata* in the United States it should be noted that the Colorado potato beetle is found extensively in Mexico where it is collected mainly on *S. rostratum* and *S. angustifolium*.

Specimens examined.—(Fig. 38, map) 349: ARIZONA: 2, Douglas, Sept. 4, 1933 (CISC); 8, 10 mi. N. Sonoita, Aug. 26, 1950 (AMNH); 2, Nogales, June 14, 1968 (USNM); ARKANSAS: 1, Jonesboro, March 8, 1965 (ISUC); COLORADO: 5, La Junta, Aug. 12, 1920 (AMNH); 1, Denver, July 12, 1938 (CASC): IDAHO: 4 Moscow, June 2, 1937 (ISUC); ILLINOIS: 2, Urbana, July 10, 1928 (UCDC); INDIANA: 3, Lafayette, May 13, 1932 (PURC); 39, Culver, July 15-17, 1969 (RLJC); 1, Posey Co., March 23, 1904 (PURC); 1, Lake Co., June 10, 1898 (PURC); 1, Vigo Col., June 10, 1898 (PURC); IOWA: 6, Ames, July 1936 (ISUC); KANSAS: 27, Peabody, July 11, 1946 (UCDC); MASSACHUSETTS: 2, Arlington, Aug. 1927 (ISUC); MINNESOTA: 2, Duluth (LACM); 1, Minneapolis, June 3, 1930 (LACM); MISSISSIPPI: 9, Ashland, Sept. 3, 1934 (UCDC); NEBRASKA: 1, Atlanta, July 23, 1966 (TAMC); NEW MEXICO: 1, Carlsbad, June 20, 1947 (UCDC); NEW HAMPSHIRE: 1, Laconia, June 1961 (CASC); NEW YORK: 40, Schenectady Co., Sept. 22, 1971 (RLJC); 2, Batavia, Aug. 28, 1913 (ISUC); NORTH CAROLINA: 1, Raleigh, July 11, 1950 (CISC); OHIO: 4, Dayton, Aug. 25, 1963 (ISUC); 1, Canton, Aug. 10, 1937 (CASC); PENNSYLVANIA: 1, Easton, July 1, 1913 (CASC); OKLAHOMA: 2, Ft. Sill, May 9, 1954 (ISUC); 4, Willis, May 29, 1961 (ISUC); SOUTH DAKOTA: 1, Hot Springs, Sept. 19, 1906 (CASC); TENNESSEE: L, Nashville, Aug. 4, 1897 (CASC); TEXAS: 1, San Angelo, June 28, 1948 (AMNH); 1, Ft. Davis, June 25, 1942, (CASC), WASHINGTON: 4, Soap Lake, Sept. 8, 1960 (ISUC); 6, Pullman, Sept. 14, 1967 (ISUC); WISCONSIN: 1, Winnebago Co. June 5, 1936 (ISUC); WYOMING: 1, 33 m. S. Newcastle, Sept. 10, 1949 (AMNH); CANADA: BRITISH COLUMBIA: 2, Fernie, June 4, 1936 (CASC); QUEBEC: 1, Montreal (CASC); MEXICO: CHIHUAHUA: 4, Salaices, Aug. 20, 1947 (AMNH); COLIMA: 7, 12 mi. SW. Colima. July 20, 1966 (TAMC); DISTRITO FEDERAL: 1, Mexico City (AMNH);

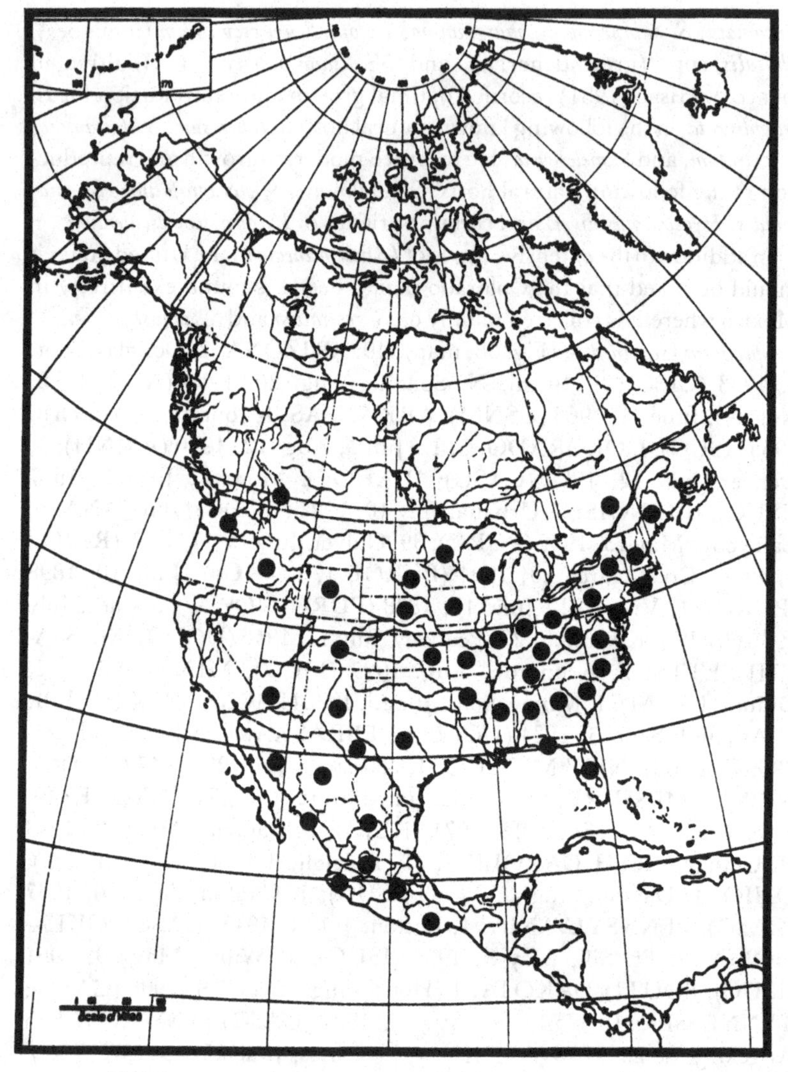

Fig. 38. Map of the distribution of *L. decemlineata* (Say) (circle).

DURANGO: 9, Durango, Aug. 1, 1964 (CISC); 2, Durango, Aug. 22, 1953 (AMNH); JALISCO: 7, Guadalajara, July 21, 1965 (OSUC); 4, Lagos de Moreno, Aug. 10, 1962 (CISC); MICHOACAN: 1, 4 mi. E. Apatzingan, July 21, 1954 (CISC); 1, 11 mi. E. Apatzingan, Aug. 20, 1954 (CISC); MORELOS: 2, 2 mi. S. Cuernevaca, July 8, 1962 (CISC); NAYARIT: 1, Tuxpan, June 29, 1962 (CISC); OAXACA: 7, Oaxaca,

Aug. 16, 1903 (AMNH); PUEBLA: 3, 11 mi. SE. Acatlan, July 11, 1952 (CISC); QUERETARO: 1, Queretaro, Aug. 9, 1962 (CISC); SINALOA: 12, 12 mi. N. Mazatan, July 26, 1965 (OSUC); 21, Los Moch's, Aug. 17, 1922 (CASC); 3, 10 mi. NW. Maxatlan, Aug. 15, 1965 (TAMC); 17, Elota, June 27, 1962 (CISC); SONORA: 13, Alamos, Oct. 10, 1952 (AMNH); 40, 10 mi. W. Alamos, July 21, 1954 (AMNH); ZACATECAS: 17 mi. N. Fresnillo, July 16, 1962 (CISC).

Discussion.—Specimens examined from Mexico and southwestern United States tend to be darker, especially on the ventral side, but not to the extent of *L. undecemlineata* which is entirely black on the ventral side including the legs. Most records of *L. decemlineata* come from the western section of Mexico. In the eastern sections of Mexico, *L. undecemlineata* is common but the ranges of both species do overlap.

Stal (1859) described *L. multitaeniata* as an intermediate species between *L. decemlineata* and *L. undecemlineata*. Jacoby (1883) considered *L. multitaeniata* as a variety of *L. decemlineata*. This species is here made a synonym of *L. decemlineata*.

Tower (1906) described 9 varieties of *L. decemlineata* including: *tortuosa, pallida, defectopunctata, melanicum, minuta, immaculothorax, albida, rubrivittata,* and *obscurata*. He also described 3 species in 1906 including: *L. intermedia, L. oblongata,* and *L. rubicunda*. After careful study of *L. decemlineata* and specimens labeled by Tower the 3 Tower (1906) species were made synonyms of *L. decemlineata*. Studies made on the male genitalia assisted in this conclusion.

Leptinotarsa chalcospila Stal

Leptinotarsa chalcospila Stal, 1858. Ofv. Svenska Vet.-Akad. Forh. 15:476.
Type locality: Mexico

Stal, 1863:157; Tower, 1906:5; Blackwelder, 1946:673; Jacoby, 1883:227.

The immaculate, aeneous head, thorax, and scutellum; head and thorax punctation fine; chestnut brown elytra, each elytron with 7-10 black spots, large, 14-16 mm in length and 10-11 mm in width are characters which will separate this species from all others in North America.

Head: immaculate, dark aeneous; punctation fine, dense; interocular distance 2.4 mm, head width 3.0 mm.

Thorax: immaculate, dark aeneous; punctation fine, courser near lateral margins, dense; anterior-lateral angles pointed; pronotum length 3.1 mm, pronotum width 7.6 mm. Scutellum dark aeneous; punctation fine, dense. Elytra (Fig. 39) oval, very convex, chestnut brown, margin black,

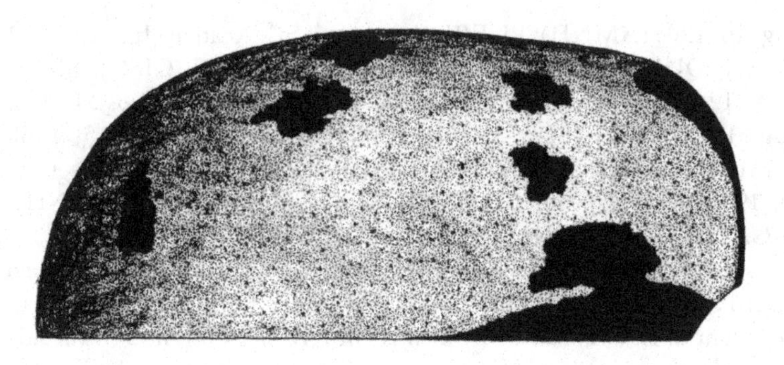

Fig. 39. Left elytron of *L. chalcospila* Stal, dorsal view.

each elytron with black irregular spot at humeral angle, large black expanded area at sutural margin just below scutellum; 3 black spots at base of each elytron but below humeral angle spot; 3 black spots at center of each elytron, spot closest sutural margin almost forming 2 spots; 2 black spots and a black dash line located at apex of elytron; punctation coarse throughout, dense; elytra length 11.4 mm, elytra width 10.1 mm.

Legs: black, unicolorous.

Abdomen: black, unicolorous, dense fine white pubescence.

Individual variation.—Only 1 specimen examined, total length 14.5 mm, greatest width 10.1 mm, interocular distance 2.4 mm, head width 3.0 mm, pronotum length 3.1 mm, pronotum width 7.6 mm, elytra length 11.4 mm, elytra width 10.1 mm. Stal (1858), in his original description, described this species as 14-16 mm in length and 10-11 mm in width.

Specimens examined.—1: MEXICO: 1, (USNM).

Discussion.—Jacoby (1883) reports this species as very rare and he examined only 3 specimens, 1 in the Salle collection and 2 from the Baly collection.

Jacoby (1891) reports this species in Tacambaro de Codallos and Huetamo de Nunez both in the state of Michoacan, Mexico. Tower (1906) gives only Mexico as the locality for this species and includes no discussion.

This is one of the largest species of *Leptinotarsa*. The chestnut brown elytra with black spots distinguishes this species quite easily.

Leptinotarsa lacerata Stal

Leptinotarsa hopferni Dejean, 1836. Catalogue des Coleopteres de le collection de M. le comte Dejean. 4:421. (*Nom. Nud.*).

Type locality: Mexico

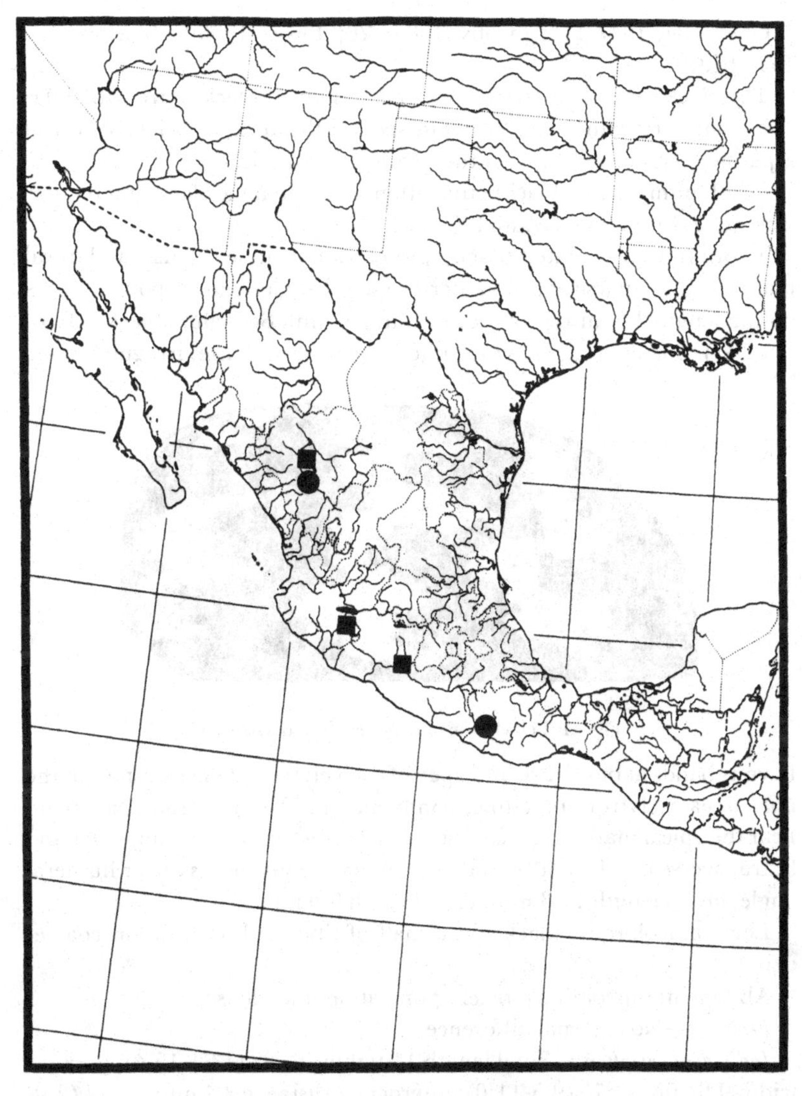

Fig. 40. Map of the distribution of *L. chalcospila* Stal (square), and *L. lacerata* Stal (circle).

Leptinotarsa lacerata Stal, 1858. Ofv. Svenska Vet.-Akad. Forh. 15:476. (Synonymy, Jacoby 1883:228.)
Type locality: Mexico.

Chrysomela lacerata (Stal, 1863). Monographie des Chrysomelides de l'Amerique. 4(3):157. (New generic assignment.)

Chevrolat, 1877:143; Jacoby, 1883:228; Tower, 1906:5; Blackwelder, 1946:673.

The black head, pronotum, legs, and abdomen, black elytra with 3 flavous transverse bands and 2 flavous spots at the apex of each elytron will separate this species from all others in North America.

Head: immaculate, black; punctation fine; antennae black; interocular distance 2.1 mm, head width 2.6 mm.

Thorax: immaculate, black; punctation course, dense at lateral margins, fine, sparse at disk; anterior-lateral angles bluntly pointed; pronotum length 2.5 mm, pronotum width 7.0 mm. Scutellum black. Elytra (Fig. 41) broadly oval, convex; punctation course, dense, irregular rows,

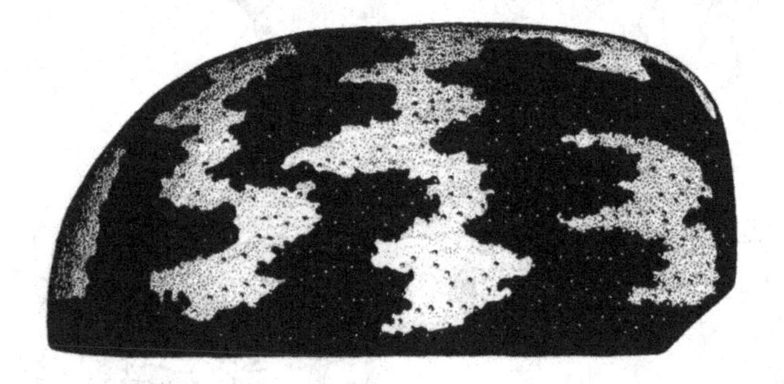

Fig. 41. Left elytron of *L. lacerata* Stal, dorsal view.

flavous bands at base, disk and apex of each elytron; 2 flavous spots at the apex of each elytron sometimes confluent with the third transverse band near the apical margin, transverse band 2 sometimes confluent with band 3, transverse band 1 and 2 confluent along the side of disk near humeral angle; elytra length 11.3 mm, elytra width 9.9 mm.

Legs: unicolorous, black; distal half of tibia with punctation coarse, dense.

Abdomen: unicolorous, black, punctation fine, dense.

Females.—No external difference.

Individual variation.—Total length 14.0 mm ± .88 (13.1-15.4); greatest width 9.9 mm ± .70 (9.5-11.0); interocular distance 2.1 mm ± .14 (2.0-2.3); head width 2.6 mm ± .08 (2.5-2.7); pronotum length 2.6 mm ± .10 (2.5-2.7); pronotum width 7.0 mm ± .26 (6.7-7.3); elytra length 11.3 mm ± .63 (10.5-12.0); elytra width 9.9 mm ± .70 (9.5-11.0).

The transverse flavous bands will vary in thickness and pattern slightly. I consider black as the base color while Jacoby (1883) considers the flavous color as the base. Black usually predominates the elytra.

Specimens examined.—(Fig. 40, map) 5: MEXICO: 4 (USNM); 1, OAXACA: 1, Monte Alban, Oct. 12, 1963 (USNM).

Discussion.—This species is reported by Jacoby (1883) as common, yet only five specimens were available for study. Tower (1906) reported this species in northern and eastern part of Oaxaca and in the state of Chiapas. Additional localities recorded are Veracruz and Morelos.

Leptinotarsa heydeni Stal

Leptinotarsa heydenii Hopfner, 1836. Dejean catalogue des Coleopteres de la collection de M. le comte Dejean. 4:421. (*Nom. Nud.*)
Type locality: Mexico.

Leptinotarsa heydeni Stal, 1858. Ofv. Svenska Vet.-Akad. 15:475. =
Type locality: Mexico

Chrysomela heydeni (Stal, 1858). Monographie des Chrysomelides l'Amerique. 4(3):158, 1863. (New generic assignment.)

Motschulsky, 1860:182; Chevrolat, 1877:143; Jacoby, 1883:228; Tower, 1906:5; Blackwelder, 1946:673.

Distinctive features: the dark blue to black head and pronotum; 7 flavous spots on each elytron; the middle spot near the sutural margin is sometimes confluent with the elongated spot near lateral margin forming a transverse band, base color always metallic blue.

Head: dark blue to black, unicolorous; punctation fine, dense; antennae unicolorous, brown, distal 5 segments expanded, fine pubescence; interocular distance 2.0 mm, head width 2.5 mm.

Thorax: immaculate, black; punctation course at lateral margins, fine at disk; anterior-lateral angles bluntly pointed; pronotum length 2.4 mm, pronotum width 6.6 mm. Scutellum blue to black, punctation fine, sparse. Elytra (Fig. 42) broadly oval, convex; flavous spots on each elytron, the

Fig. 42. Left elytron of *L. heydeni* stal, dorsal view.

middle spot nearest the sutural margin is sometimes confluent with the elongated spot near the lateral margin forming a transverse band; punctation coarse, dense, sometimes in irregular rows; elytra length 10.0 mm, elytra width 9.2 mm.

Legs: unicolorous, metallic blue; femur punctation sparse; punctation on distal half of tibia, coarse, dense.

Females.—No apparent difference.

Individual variation.—Total length 12.8 mm ± .56 (12.0-13.9); greatest width 9.2 mm ± .43 (8.6-10.0); interocular distance 2.0 mm ± .12 (1.9-2.5); head width 2.5 mm ± .15 (2.3-2.7); pronotum length 2.4 mm ± .17 (2.1-2.7); pronotum width 6.6 mm ± .22 (6.4-7.1); elytra length 10.0 mm ± .62 (8.6-11.5); elytra width 9.2 mm ± .43 (8.6-10.0).

The base color of the elytra may appear more black but the usual color is blue. The middle flavous spot varies in some specimens, and sometimes forms a transverse band with the elongated spot on the lateral margin.

Specimens examined.—(Fig. 43, map) 26: MEXICO: 11, (USNM); 2, (INHS); DURANGO: 2, Sierra de Durange (USNM); PUEBLA: 1, Huauchinango (USNM); VERACRUZ: 5, Tolome near Rinconada, July 27, 1955 (AMNH); 5, Rincanada, June 19, 1951 (CISC).

Stal (1858), in his original description, recorded the type locality as Mexico, but later Stal (1863) recorded the locality as Brazil. Jacoby (1883) doubts this second locality. Specimens studied were mainly from the Veracruz area. Tower (1906) adds the following localities: Taneza and Almolonga in Oaxaca and Tremax in northern Yucatan. Hsiao and Hsiao (1983) reported collecting adults and pupa of *L. heydeni* on *Viguiera dentata* (Cav.) Sprengel at Uman, Yucatan, Mexico.

This species is the type species of the genus, designated by Motschulsky (1860).

Leptinotarsa novemlineata Stal

Leptinotarsa novemlineata Stal, 1860. Ofv. Svenska Vet.-Akad. Forh. 17:456.

Type locality: Mexico

Chrysomela novemlineata (Stal, 1863) Monographie des Chrysomelides de l'Amerique. 2:159 (New generic assignment.)

Jacoby, 1883:230; Tower, 1906:8; Blackwelder, 1946:673.

Distinctive features: head and pronotum bronze to coppery-green; maxillary palpi and antennae yellow-brown; elytra yellow-brown with 4 distinct vittae on each elytron, vittae brown; thorax and elytra strongly punctured, with a transverse spot sometimes found in the middle of the

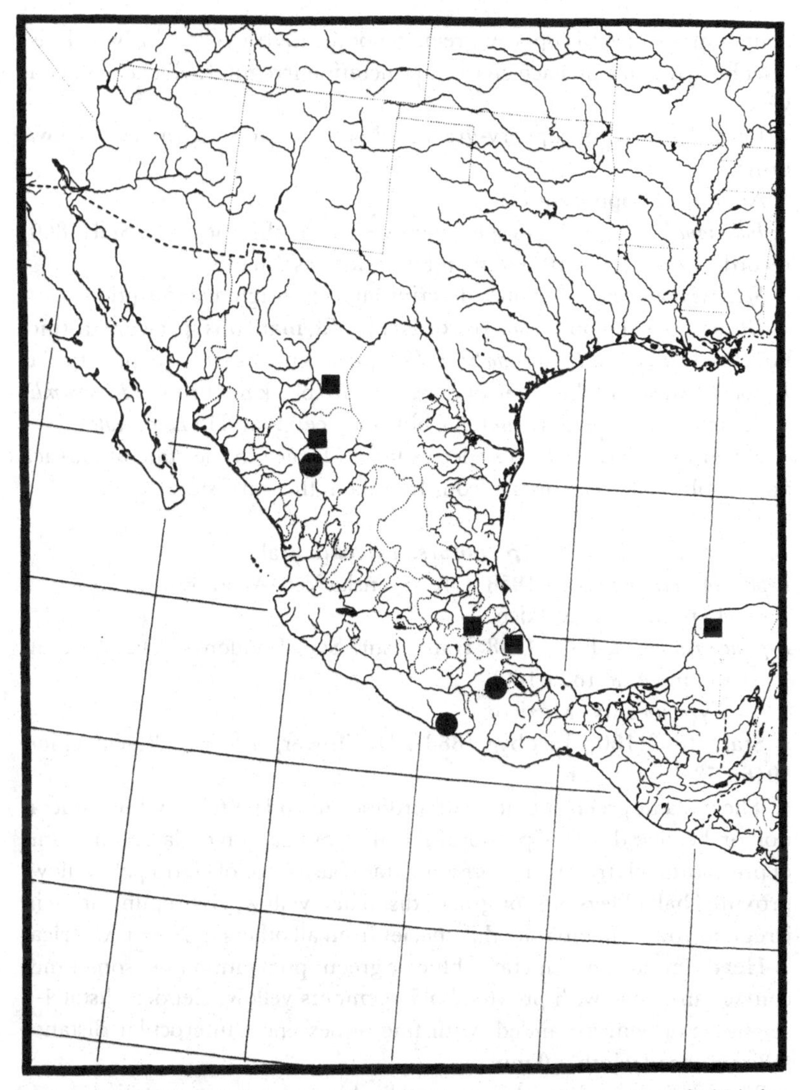

Fig. 43. Map of the distribution of *L. heydeni* Stal (square), and *L. novemlineata* Stal (circle).

elytra between the first and second vittae; ventral surface coppery; legs coppery with tibia and tarsal segments yellow brown.

Head: unicolorous, bronze to coppery-green; punctation distinct; maxillary palpi and antennae yellow-brown.

Thorax: unicolorous, bronze to coppery-green; punctation distinct.

Scutellum bronze to coppery-green, smooth. Elytra yellow brown, 4 distinct brown vittae on each elytron; punctation irregular; elytra oval, convex.

Legs: bronze to coppery-green; tibia and tarsal segments yellow-brown.

Abdomen: coppery-green.

Individual variation.—No specimens were available for study. Stal (1860) records the length as 10.0 mm and the width as 7.0 mm.

Specimens examined.—None, description from original description.

Discussion.—Jacoby (1883) records (Fig. 43, map) this species as distinct but closely allied to *L. calceata* Stal. *L. calceata* is not as strongly punctate as *L. novemlineata* and the vittae of *L. calceata* are black not brown. *L. novemlineata* is more bronze than metallic blue or green found in *L. calceata.*

Tower (1906) records this species from Oaxaca in the state of Oaxaca and Jacoby (1883) records it from Juquila in the same state.

Leptinotarsa calceata Stal

Leptinotarsa calceata Stal, 1858. Ofv. Svenska Vet.-Akad. Forh. 15:476.
Type locality: Mexico.

Leptinotarsa vittata Baly, 1858 Trans. Ent. Soc. London. 4(2):351. (Synonymy, Stal 1863:160).
Type locality: Mexico.

Stal, 1863:160; Jacoby, 1883:231; Tower, 1906:7; Blackwelder, 1946:673.

The metallic green or blue head, pronotum, and scutellum, fine punctation on head and disk of pronotum, course punctation on lateral margins of pronotum, elytron with 4 brown vittae, base color of elytra pale yellow, proximal half of legs blue or green, distal half yellow, elytra punctation in irregular rows will separate this species from all others in North America.

Head: immaculate, metallic blue or green; punctation fine, sometimes course; antennae with proximal 5-7 segments yellow, slender, distal 4-6 segments brown, expanded, with fine pubescence; interocular distance 1.8 mm, head width 2.0 mm.

Thorax: immaculate, shinny, metallic blue or green; punctation variable on pronotum; anterior-lateral margins of pronotum rounded; pronotum length 2.0 mm, pronotum width 5.1 mm. Scutellum unicolorous, shinny, brown, green or blue, smooth. Elytra (Fig. 44) oval, convex; punctation course, irregular rows extending from base to apex of elytron pale yellow with 4 brown vittae extending from base to apex of elytron; sutural margin brown; elytra length 7.6 mm, elytra width 6.8 mm.

Fig. 44. Left elytron of *L. calceata* Stal, dorsal view.

Legs: coxae, femur, and proximal half of tibia metallic blue or green punctation course, sparse; distal half of tibia and tarsus pale yellow to golden.

Abdomen: unicolorous, metallic blue or green, punctation fine, scant pubescence.

Females.—No external difference.

Individual variation.—Total length 10.0 mm ± .87 (8.6-12.2); greatest width 6.9 mm ± .61 (6.0-8.3); interocular distance 1.8 mm ± .18 (1.5-2.0); head width 2.0 mm ± .16 (1.7-2.3); pronotum length 2.0 mm ± .14 (1.8-2.4); pronotum width 5.1 mm ± .33 (4.7-5.7); elytra length 7.6 mm ± .66 (6.1-8.7); elytra width 6.9 mm ± .61 (6.0-8.3).

Logan *et. al.* (1983) records *L. calceata* at Alvarado bridge in Vera Cruz in early September feeding on a Composite. It should be noted that a number of *Leptinotarsa* including *L. cacica*, and *L. lineolata* feed on plants in the family Compositae.

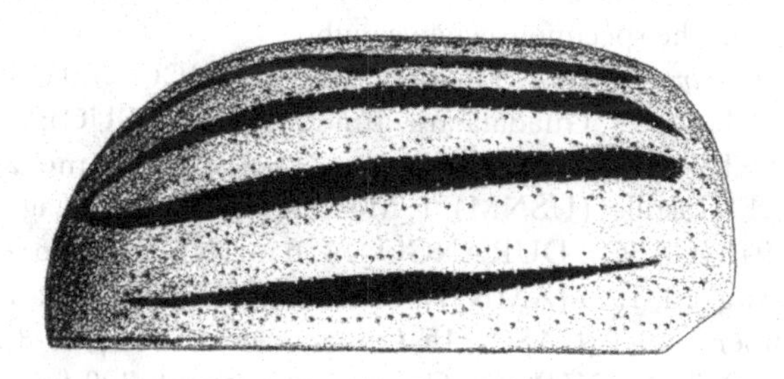

Fig. 46. Left elytron of *L. melanothorax* (Stal), dorsal view.

Leptinotarsa melanothorax (Stal)

Myocoryna melanothorax Stal, 1859. Ofv. Svenska Vet.-Akad. Forh. 16:317.
Type locality: Mexico

Chrysomela melanothorax (Stal, 1863), Monographie des Chrysomelides de l'Amerique. 2:166 (New generic assignment.)

Leptinotarsa melanothorax Jacoby, 1883. Biologia Centrali-Americana, Insecta, Coleoptera. 6(1):234. (New generic assignment.)

Horn, 1884:128; Linell, 1896:196; Tower, 1906:7; Leng, 1920:294; Blackwelder, 1946:673.

Distinctive features: black head and thorax; elytra pale yellow, each elytron with 4 dark brown vittae; abdomen and legs black.

Head: immaculate, black; mandibles, maxillary palps, antennae black; punctation sparse, fine; interocular distance 1.7 mm, head width 2.0 mm.

Thorax: immaculate, black; punctation fine, sparse; anterior-lateral angles rounded; pronotum length 2.3 mm, pronotum width 5.0 mm. Scutellum black, smooth, Elytra (Fig. 46) oval, convex; each elytron pale yellow with 4 dark brown to black vittae extending from just below the base of the elytra and sometimes joining at the apex; a faint line runs adjacent to the elytra margin; punctation course, in irregular rows bordering each vittae; elytra length 7.1 mm, elytra width 7.3 mm.

Legs: unicolorous, black.

Abdomen: immaculate, black.

Females.—No external difference.

Individual variation.—Total length 9.4 mm ± .47 (8.4-10.2); greatest width 7.0 mm ± .40 (6.5-7.7); interocular distance 1.7 mm ± .11 (1.5-1.9); head width 2.0 mm ± .12 (1.8-2.3); pronotum length 2.3 mm ± .18 (2.0-2.6); pronotum width 5.0 mm ± .29 (4.6-5.5); elytra length 7.1 mm ± .58 (6.4-8.3); elytra width 7.0 mm ± .40 (6.5-7.7).

The vittae may join at the apex of the elytron. Some specimens appear darker than others, even in the same series. This is probably caused by the different ages of the specimens when caught.

Specimens examined. —(Fig. 45, map) 127; MEXICO: 3, (USNM); 2, (CASC); JALISCO: 2, Guadalajara, June 24, 1903 (OSUC); 8, Guadalajara, June-July, 1903 (AMNH); 1, Zapotlanejo, June 24, 1903 (OSUC); 1, El Castillo, (USNM); 1, Mascota (USNM); 1, Lago de Chapala, July 1940 (CASC); DURANGO: 14, 6 mi. NE. El Salto, elevation 8,500 feet, Aug. 10, 1947 (AMNH); 4, 8 mi. N. Coyotes, elevation 7,500 feet, November 7, 1970 (CASC); 15, Otinapa, elevation 7,500-8,200 feet, Aug. 7-111, 1947 (AMNH); 17, Coyotes, elevation 8,300 feet, Aug. 8, 1947 (AMNH); 53, Palos Colorados, Aug. 5, 1947 (AMNH); YUCATAN; 1, (USNM); DISTRITO FEDERAL: 2, (USNM); PUEBLA: 1, Huauchinango (USNM).

Discussion.—Jacoby (1883) reports this species in Toluca, Puebla and Guanajuato. Tower (1906) examined specimens collected by Hoge, Salle,

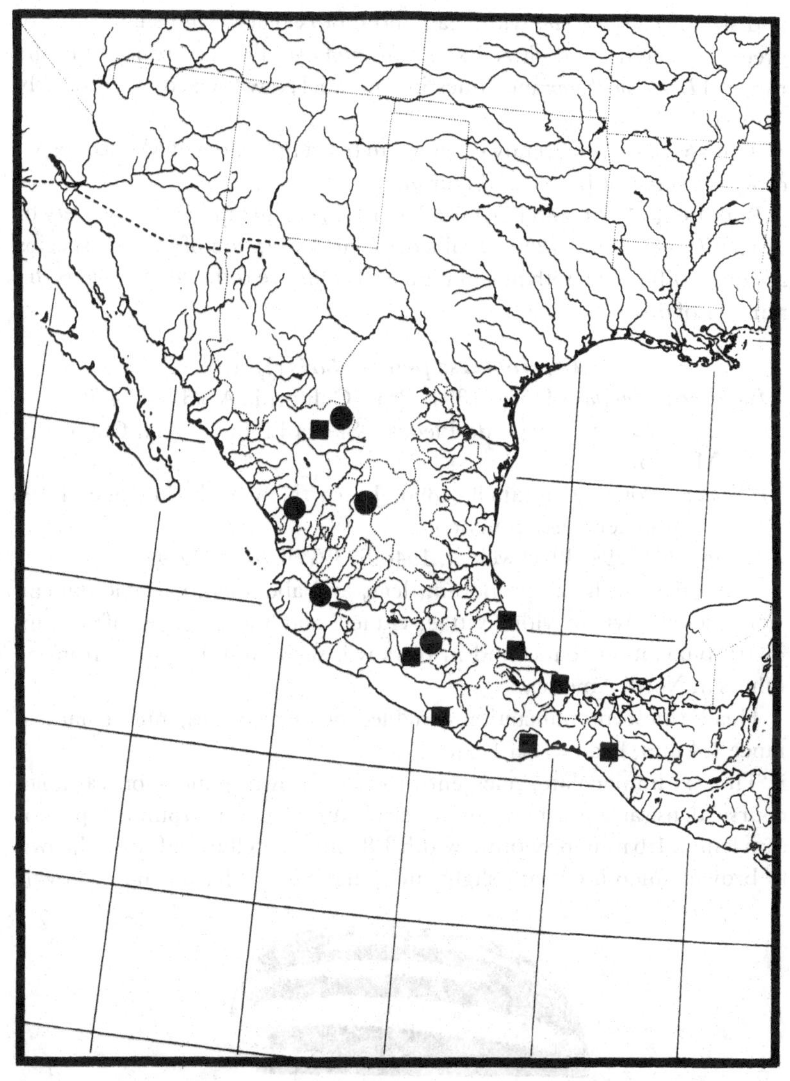

Fig. 45. Map of the distribution of *L. calceata* Stal (square), and *L. melanothorax* (Stal) (circle).

and Duges in Mexico City and Guadulupe in the Federal District; Toluca, Puebla, Guanajuato, and Morelia.

Horn (1884), in working over specimens in LeConte's collection, tells of a single specimen collected by Prof. Snow in New Mexico and identified by LeConte as *Doryphora melanothorax* Stal. Horn remarks that it is the form and size of *L. haldemani* with head and thorax black, with slight tinge of green, and elytra and vittae like *L. decemlineata*. This record is out of the range of *L. melanothorax* and I question the label New Mexico, it is probably Mexico.

This species is apparently confined to the central area of Mexico, especially the states of Jalisco and Durango.

Both Leng (1920) and Blackwelder (1946) list this species as a variety of *L. multitaeniata* (Stal, 1859). Stal's original description of the two species are quite different. Both maculation and color are different as well as the number of distinct vittae.

Leptinotarsa peninsularis (Horn)

Myocoryna peninsularis Horn, 1894. Proc. California Acad. Sci. 4:407.
Type locality: corral de Piedras, Sierra El Tasta, Baja California, Mexico.

Leptinotarsa peninsularis Linell, 1896. Journ. New York Ent. Soc. 4:196 (New generic assignment.)

Leng, 1920:294; Blackwelder, 1946:673; Wilcox, 1972:9.

The rufescent head, pronotum, legs, and abdomen, variable punctation, flavous elytron with 3 vittae, vittae 2 and 3 join at apex of elytron, elytra punctation coarse and in rows will separate this species from all others in North America.

Head: rufescent; punctation variable, course near eyes; interocular distance 1.4 mm, head width 1.6 mm.

Thorax: immaculate, refescent-brown to brown; punctation variable, course, dense at lateral margins; anterior-lateral margins rounded; pronotum length 1.6 mm, pronotum width 3.8 mm. Scutellum rufescent-brown to brown; smooth or with slight fine punctation. Elytra (Fig. 47) oval,

Fig. 47. Left elytron of *L. peninsularis* (Horn), dorsal view.

slightly elongated, convex; base color of elytron flavous, sutural margin and lateral margins of elytron brown tapering at the apex and not joining, vittae originating about 1 mm below the basal margin, vittae brown, vittae 1 joins with sutural margin at upper quarter, vittae 2 and 3 joining at apex of elytron into a point, but it does not reach the edge of the elytron, base of vittae 3 expanded to twice its width; punctation course, in regular rows outlining each of the vittae; expanded part of vittae 3 was probably the base of vitta 4 since rows of punctation extend from this area outlining a vitta but no color is present between the punctation; there are 9 rows of punctures on each elytron; elytra length 5.7 mm, width 5.0 mm.

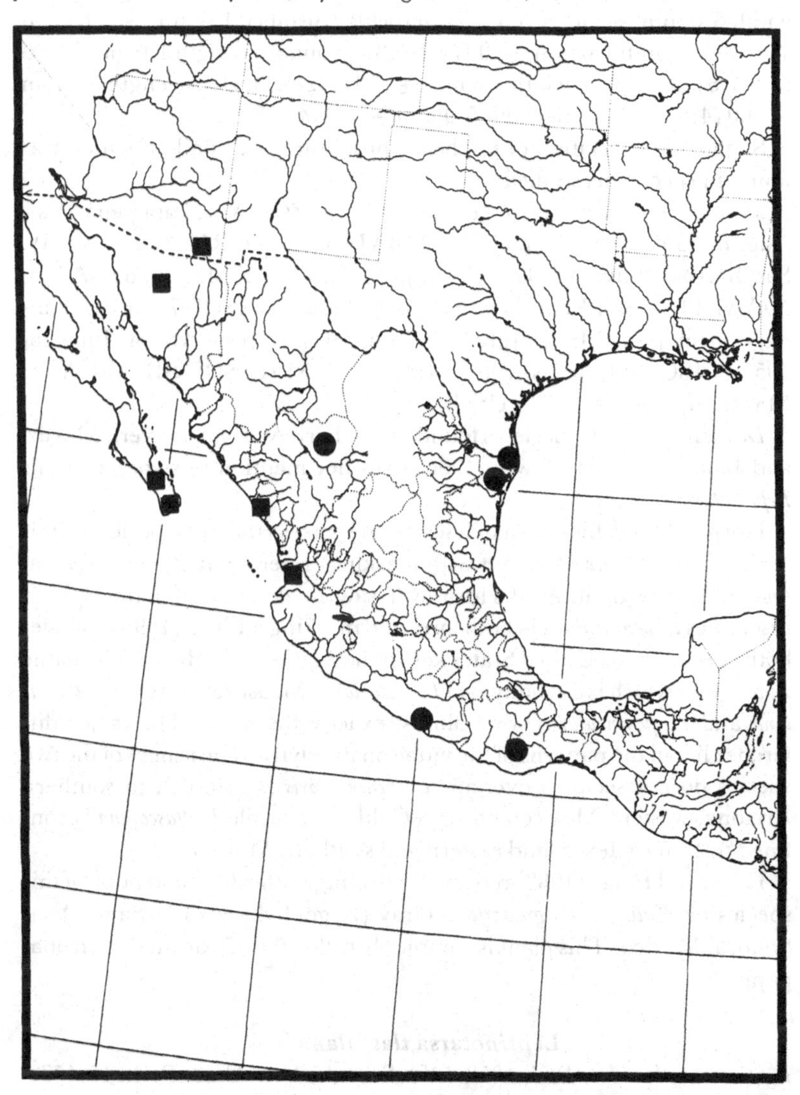

Fig. 48. Map of the distribution of *L. peninsularis* (Horn) (square), and *L. tlascalana* Stal (circle).

Legs: unicolorous, rufescent to brown.

Abdomen: unicolorous, rufescent to brown; punctation sparse; pubescent fine.

Females.—No external difference.

Individual variation.—Total length 7.4 mm ± .45 (7.1-8.1); greatest width 5.0 mm ± .30 (4.5-5.5); interocular distance 1.4 mm ± .06 (1.3-1.5); head width 1.6 mm ± .08 (1.4-1.8); pronotum length 1.6 mm ± .09 (1.5-1.8); pronotum width 3.8 mm ± .23 (3.2-4.1); elytra length 5.7 mm ± .60 (4.5-6.5); elytra width 5.0 mm ± .30 (4.5-5.5).

Slight color variation of the head, pronotum, legs and abdomen was found in specimens studied.

Specimens examined.—(Fig. 48, map) 33. ARIZONA: 1, Patagonia Mts., Aug. 8, 1952 (OSUC); MEXICO: BAJA CALIFORNIA: 1, 5 mi. W. San Bartolo, July 13, 1938 (CASC); 2, Trunfo, July 13, 1938 (CASC); NAYARIT: 1, San Juan, Peyotan, Aug. 2, 1955 (CISC); 7, Arroyo Santiago, near Jesus Maria, July 5, 1955 (CISC); 6, Jesus Maria, June 26, 1955 (CISC); 14, Jesus Maria, July 27, 1955 (CISC); SINALOA: 1, Mazatlan, Aug. 10, 1970 (USUC).

Discussion.—This species is found in southern Arizona, western Mexico and Baja California. Tower (1906) does not mention this species in his *Leptinotarsa* studies.

Horn (1894) in his original description remarks that this species is similar in form to *L. lineolata*, but with markings resembling *Zygogramma contina*. It is very doubtful if Horn ever saw *L. tlascalana*, for this species resembles *L. peninsularis* in both form and markings. Linell (1896) includes both species in his United States key to the *Leptinotarsa*. He uses the name *L. dahlbomi* which is a synonym of *L. tlascalana*. Measurements for *L. tlascalana* and *L. peninsularsis* were almost exactly the same. The major difference lies in the patterns of the vittae on the elytra. The ranges of the two species do not seem to overlap. *L. peninsularis* is confined to southern Arizona, western Mexico and Baja California, while *L. tlascalana* is confined to eastern Texas, and eastern and southern Mexico.

Hsiao and Hsiao (1983) reported collecting both adults and pupa of this species on *Kallstroemia grandiflora* Gray (Zygophyllaceae) in Santa Ana, Sonora, Mexico. This plant is commonly called the Mexican, or Arizona, poppy.

Leptinotarsa tlascalana Stal

Leptinotarsa tlascalana Stal, 1858. Ofv. Svenska Vet.-Akad. Forh. 15:477.
Type locality: Mexico.

Myocoryna dahlbomi Stal, 1859. Ofv. Svenska Vet.-Akad. Forh. 16:317. (NEW SYNONYMY.)

Type locality: Mexico and Texas.

Chrysomela tlascalana (Stal, 1863) Monographie des Chrysomelides de l'Amerique. 2:158 (New generic assignment.)

Crotch, 1873:47; Jacoby, 1883:239; Linell, 1896:5; Leng, 1920:294; Blackwelder, 1946:673.

Distinctive features: brown to black head and pronotum; elytra dark brown to black, each elytron with two flavous vittae, one near the sutural margin, the other near the apical margin, both joining at apex of elytron; punctation in regular rows from base to apex of elytron; legs and abdomen unicolorous, light to dark brown.

Head: light brown to black; punctation course, dense near eyes; interocular distance 1.3-1.5 mm, head width 1.6-1.7 mm.

Thorax: immaculate; light brown to black; anterior-lateral margins of pronotum rounded; punctation fine, course, dense along lateral margins; pronotum length 1.5 mm, width 3.8 mm. Scutellum light brown to black, nearly smooth. Elytra (Fig. 49) oval, slightly elongated, convex; base color

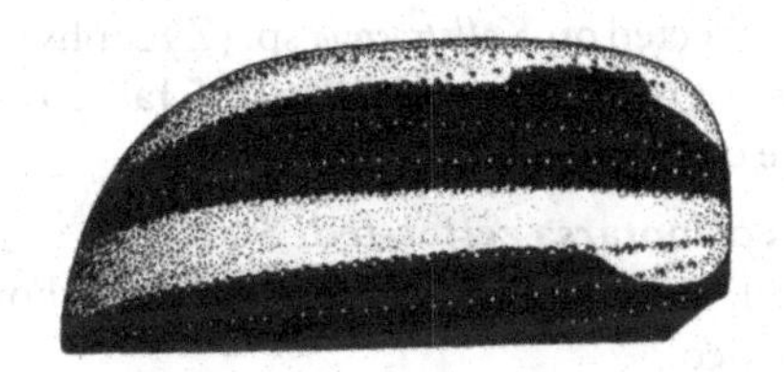

Fig. 49. Left elytron of *L. tlascalana* Stal, dorsal view.

of elytra dark brown to black; each elytron with 2 flavous vittae, vitta 2 near the sutural margin, at least .5 mm wide, extends from base to apex of elytron; vitta 2 extends from humeral angle to apex of elytron where it joins with vitta 1; vitta 2 with indentation near its base extending for at least 1 mm; punctation coarse, in regular rows extending from base to apex, usually with 9 complete punctation rows on each elytron, and rows converge at apex; length 5.6 mm, width 5.0 mm.

Legs: unicolorous, light to dark brown.

Abdomen: light to dark brown, no punctation, scant pubescence.

Females.—No external difference.

Individual variation.—Total length 7.2 mm ± .37 (6.5-7.5); greatest width 5.0 mm ± .28 (4.5-5.4); interocular distance 1.5 mm ± .07 (1.3-1.5); head width 1.6 mm ± .04 (1.6-1.7); pronotum length 1.4 mm ± .08

(1.5-1.6); pronotum width 3.8 mm ± .19 (3.5-4.0); elytra length 5.6 mm ± .39 (5.0-6.0); elytra width 5.0 mm ± .28 (4.5-5.4).

Variation is found in the overall coloration of head, pronotum, and base color of elytron from light brown to black. The width of the vittae varies but is usually near .5 mm. Elytra punctation is in regular rows but sometimes rows mix at the disk of elytron.

Specimens examined.—(Fig. 48, map) 7: TEXAS: 4, Brownsville, June 23-25, 1930 (UCDC); MEXICO; CHIAPAS: 1, 12 mi. W. Ocozocoauta, July 26, 1952 (CASC); GUERRERO: 1, Iguala (AMNH); TAMAULIPAS: 1, Metamoros, (CASC).

Discussion.—The distribution is extensive; specimens have been recorded from southern Mexico to the Texas border. Jacoby (1883) records this species in Texas; Yolos, Puebla, and Yucatan in Mexico, and in Granada, in Nicaragua. Jacoby remarked that this species is not common in Mexico. Tower (1906) also noted its wide range.

This species resembles *L. peninsularis* in both overall size and color. The pattern on the elytra is different in the two species.

Logan (1983) collected both adults and larvae of *L. tlascalana* from August 21 to September 3, 1983 in the states of Guerrero, Jalisco, Oaxaca, and Morelos. All were collected on *Kallstroemia* sp. (Zygophyllaceae) probably *K. rosea*, a common ground cover in parts of Jalisco. At Barre de Navidad 82 adults were collected on this plant.

Leptinotarsa zetterstedi Stal

Leptinotarsa zetterstedti Stal, 1859. Ofv. Svenska Vet.-Akad. Forh. 16:316.
Type locality: Mexico.

Chrysomela zetterstedti (Stal. 1863). Monographie des Chrysomelides de l'Amerique. 4(3):154 (New generic assignment.)

Jacoby, 1883:229; Tower, 1906:8: Blackwelder, 1946:673.

The rufescent pronotum and head, rufescent elytra with pale yellow markings outlined in black, some of the markings confluent, abdomen and legs rufescent will separate this species from all others in North America.

Head: punctation fine, courser in area around eyes; antennae rufescent, distal 5 segments darker, expanded with fine pubescence; interocular distance 1.9 mm, head width 2.3 mm.

Thorax: immaculate, rufescent with some darker red-brown areas; punctation fine, courser in area of lateral margins. Pronotum length 2.2 mm, pronotum width 5.3 mm. Scutellum rufescent, smooth. Elytra (Fig. 50) oval, convex, punctation coarse, in irregular rows near the sutural margin, scattered in other areas; base color of elytra rufescent with flavous markings outlined in black; first flavous mark near the base of the elytron

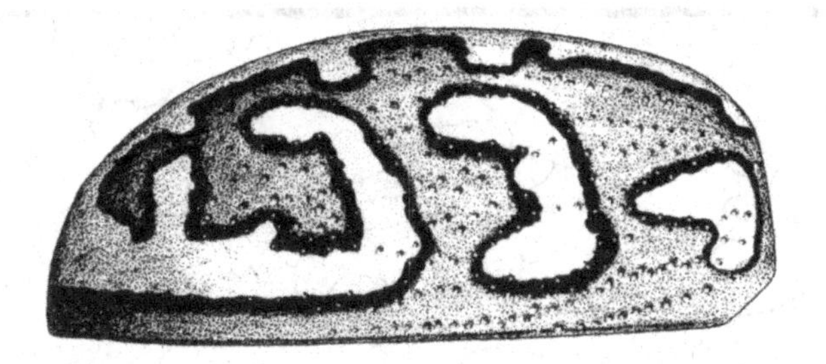

Fig. 50. Left elytron of *L. zetterstedti* (Stal), dorsal view.

elongated toward apex and sometimes confluent with second flavous mark; second marking horseshoe shaped; third marking hooked and elongated to apex where it joins apical margin markings; sometimes third marking split to form individual marks or patches; apical margin marking extends from humeral angle to apex of elytron with 4 projections extending towards disk; elytra length 8.3 mm, elytra width 7.5 mm.

Legs: unicolorous, rufescent.

Abdomen: unicolorous, rufescent; fine pubescence.

Females.—No external difference.

Individual variation.—Total length 11.0 mm ± 1.11 (9.7-12.5); greatest width 7.5 mm ± .72 (6.6-8.5); interocular distance 1.9 mm ± .10 (1.8-2.1); head width 2.3 mm ± .12 (2.2-2.5); pronotum length 2.2 mm ± .14 (2.1-2.4); pronotum width 5.3 mm ± .32 (5.0-5.8); elytra length 8.3 mm ± .90 (7.5-9.7); elytra width 7.5 mm ± .72 (6.6-8.5).

Color variation is slight, markings may be confluent along the sutural margin and in some cases nearly form a circle.

Specimens examined.—(Fig. 51, map) 6: MEXICO: JALISCO: 1, Ocotlex, Aug. 1910 (CASC); 4, 15 mi. NE. Guadalajara, Sept. 16, 1970 (USUC); ZACATECAS: 1, 10 mi. S. Jalpa, Sept. 17, 1970 (USUC).

Discussion.—Tower (1906) knew this species from Jalisco only. This species is unlikely to be confused with any others in the genus because of its unique markings.

Leptinotarsa boucardi Achard

Leptinotarsa boucardi Achard, 1923. Nouveaux Chrysomelini d'Amerique. Fragments entomolgiques 5:65.

Type locality: Mexico (Fig. 51, map).

Blackwelder, 1946:673.

Distinctive features: head, pronotum, scutellum, and legs reddish-

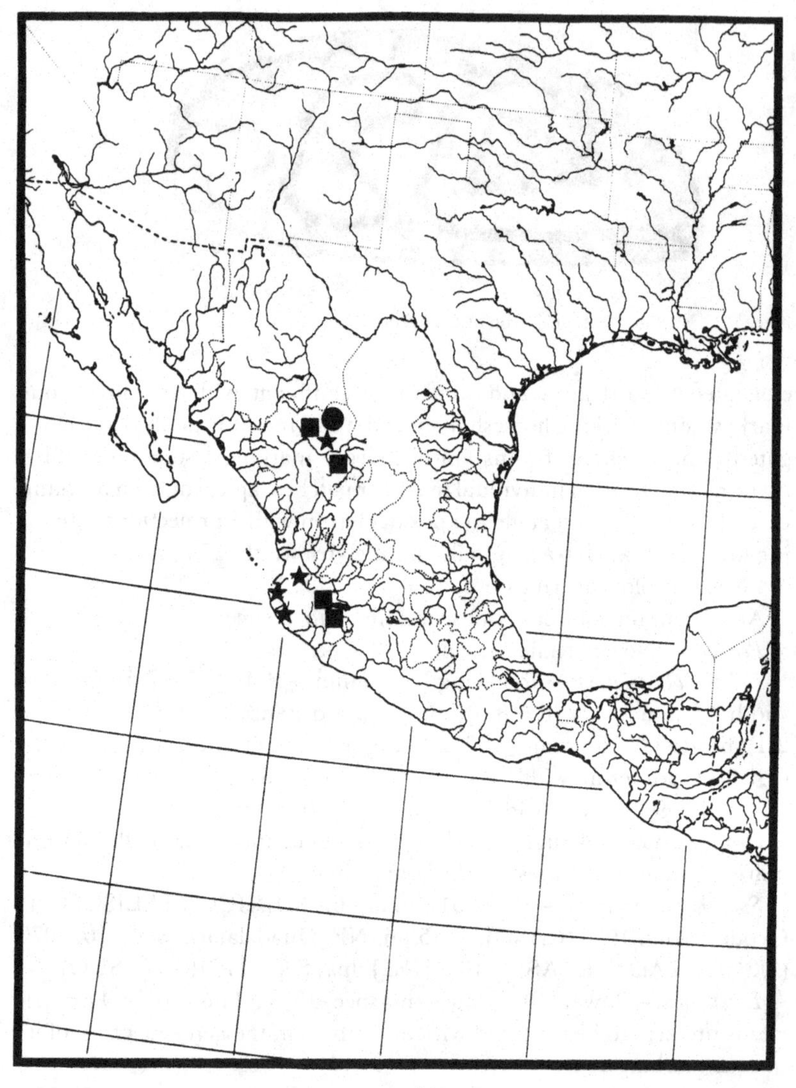

Fig. 51. Map of the distribution of *L. Zetterstedi* (Stal) (square), *L. boucardi* Achard (circle), and *L. typographica* Jacoby (star).

brown; elytra flavous, each elytron with 5 vittae; a transverse light brown line intersects each vittae; vittae 3 and 4 subdivided into fragments; ventral surface shinny blue.

Head: unicolorous, red; antennae darkened at distal end; punctation fine, irregular.

Thorax: unicolorous, reddish-brown; punctation fine, irregular, coarser at lateral margins, pronotum depressed near middle. Scutellum reddish-brown, smooth. Elytra oval, convex; base color of elytra flavous; each elytron with 5 black vittae, fifth vittae along margins of elytra, vittae 1, 2, and 5 entire; vittae 3 and 4 subdivided into fragments at apex of elytra; vittae dark brown; transverse light brown band intersects each vittae near the middle of elytra; elytra punctation dense, course, punctation absent at base of elytron between vittae 2 and 3.

Legs: unicolorous, reddish-brown.

Abdomen: shinny blue, punctation sparse.

Individual variation.—No specimens were available for study. Achard records the total length as from 9.0-10.0 mm.

Specimens examined.—None, description from original description.

Discussion.—This species is apparently not very common. Achard indicated that this species is similar to *L. typographica* Jacoby, but with elytra design similar to *L. dilecta* Stal.

Leptinotarsa typographica Jacoby

Leptinotarsa typographica Jacoby, 1891. Biologia Centrali-Americana, Insecta, Coleoptera. 6(1):254.

Type locality: Pinos Altos, state of Chihuahua, Mexico.

Blackwelder, 1946:673.

Distinctive features: flavous-red to rufescent-brown head with a black spot at the vertex, immaculate pronotum, upper half of elytron with 3 vittae, lower half spotted, some spots confluent, elytra punctation coarse, punctation surrounds the vittae and spots, ventral surface rufescent.

Head: rufescent; vertex with a black spot sometimes forms an inverted Y; punctation variable, denser in area of the eyes; apices of mandibles black; interocular distance 1.6 mm, head width 1.9 mm.

Thorax: immaculate, rufescent; punctation fine, slightly coarser near lateral margins, dense throughout; anterior-lateral margins of pronotum bluntly pointed; pronotum length 1.7 mm, pronotum width 4.6 mm. Scutellum rufescent; outlined in dark red-brown; smooth. Elytra (Fig. 52)

Fig. 52. Left elytron of *L. typographica* Jacoby, dorsal view.

oval, convex; base color of elytra flavous, basal margin of elytra black; sutural margin rufescent; vitta 1 adjacent to sutural margin extends from 1 mm below the base to the apex of the elytron, vittae 2 and 3 extend only one-third the way down the elytron from the base, a vitta 4 is curved towards the sutural margin, sometimes meeting vitta 3 forming a circle; the lower two-thirds of elytron with 12-15 major spots, some minor patches found near the apex; vittae and spots dark brown and enclosed or surrounded by course punctation; lateral margins of elytron dark brown; elytra length 6.7 mm, elytra width 6.2 mm.

Legs: unicolorous, rufescent to pale yellow.

Abdomen: rufescent to dark brown, fine pubescence.

Females.—No external difference.

Individual variation.—Total length 8.6 mm ± .50 (8.0-9.5); greatest width 6.2 mm ± .38 (5.5-6.7); interocular distance 1.6 mm ± .09 (1.5-1.7); head width 1.9 mm ± .14 (1.7-2.1); pronotum length 1.7 mm ± .17 (1.5-2.0); pronotum width 4.6 mm ± .26 (4.3-5.2); elytra length 6.7 mm ± .46 (6.1-7.5); elytra width 6.2 mm ± .38 (5.5-6.7).

Color variation from rufescent to rufescent-yellow for the head, pronotum and base color of the elytra. Vittae and spots vary in size, some spots confluent forming various patterns.

Specimens examined.—(Fig. 51, map) 42: MEXICO: JALISCO: 1, 15 mi. NE. Guadalajara, Sept. 17, 1970 (USUC); NAYARIT: 6, Tepic, June 24, 1940 (CASC); 8, Tepic, Sept. 21-24, 1953 (CASC); 22, 12 mi. SE. Tepic, Aug. 11, 1970 (USUC); 1, Tepic, Sept. 15, 1970 (USUC); 4, Compostela, Sept. 15, 1970 (USUC).

Discussion.—Jacoby (1891) and Tower (1906) remark about the restricted distribution of this species. It seems to be confined to the west-central section of Mexico in Nayarit and Jalisco.

Jacoby's original description causes some confusion by referring to the color as greenish-black. None of the specimens I have examined are in any way greenish. Tower (1906) remarks this species is closely allied to *L. lineolata*. What Tower means when he discusses the question of closely allied is sometimes questionable. The general shape of the two species are the same, but color patterns, as well as the distribution, are quite different.

Hsiao and Hsiao (1983) have collected *L. typographica* in Tepic, Nayarit, Mexico on *Lasianthaea ceanothifolia* (Compositae). Logan (1983) collected 58 adults, larvae, and eggs in late August in Tepic, Nayarit, Mexico on the same host plant.

Leptinotarsa dilecta Stal

Leptinotarsa patruelis Sturm, 1843. Catalog der Kaefer-Sammlung von Jacob Sturm, 386 pp. (*Nom. Nud.*) (Synonymy, Jacoby 1883:230.) Type locality: Mexico.

Leptinotarsa dilecta Stal, 1860. Ofv. Svenska Vet.-Akad. Forh. 17:465. Type locality: Oaxaca, state of Oaxaca, Mexico.

Chrysomela dilecta (Stal, 1860) Monographie des Chrysomelides de l'Amerique. 2:159, 1863. (New generic assignment.)

Jacoby, 1883:230; Tower, 1906:8; Blackwelder, 1946:673.

Distinctive features: head, thorax and ventral surface either green, blue or cupreous, antennae flavous, elytra base color flavous with five cupreous lines on each elytron, punctation variable and irregular, legs flavous.

Head: immaculate, metallic blue, green or cupreous; punctation fine; antennae and clypeus flavous; interocular distance 1.6 mm, head width 2.0 mm.

Thorax: metallic blue, green or cupreous; punctation fine, slightly coarser near lateral margins, anterior-lateral angles pointed, pronotum length 1.8 mm, pronotum width 5.0 mm. Scutellum metallic blue, green, or cupreous, smooth. Elytra (Fig. 53) pale yellow with Cupreous mark-

Fig. 53. Left elytron of *L. dilecta* (Stal), dorsal view.

ings; each elytron with 5 cupreous lines; sutural margin cupreous and expanded at base; line 1 adjacent and almost parallel to sutural margin; line 2 parallel to line 1; line 3 joins second at base and continues only to the disk; line 4 also joins line 2 and 3 at base of disk; line 4 also joins line 2 and 2 at base of elytron and connects with line 3 at disk; line 4 runs parallel and along the lateral margin of elytron; a transverse cupreous band extends across middle of elytron from line 1 to line 5; parts of vittae 3 and 4 found below transverse band; punctation coarse, irregular, and usually found along vittae; irregular punctation lines formed along sutural margins; elytra length 6.7 mm, elytra width 6.5 mm.

Legs: unicolorous, cupreous.

Abdomen: unicolorous, metallic blue, green, or cupreous.

Individual variation.—Only 1 specimen examined; total length 8.7 mm, greatest width 6.5mm, interocular distance 1.6 mm, head width 2.0 mm, pronotum length 1.8 mm, pronotum width 5.0 mm, elytra length 6.7 mm, elytra width 6.5 mm.

Specimens examined.—(Fig. 54, map) 1: MEXICO: 1, (USNM).

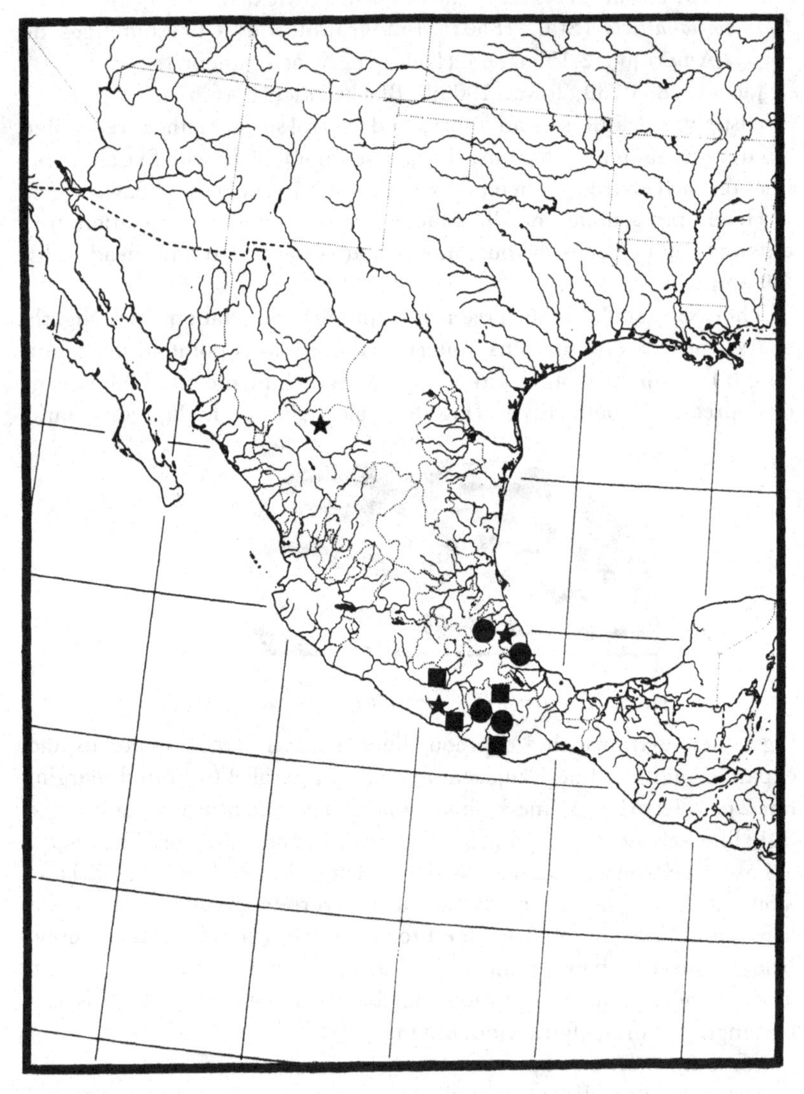

Fig. 54. Map of the distribution of *L. dilecta* (Stal) (square), *L. obliterata* (Chevrolat) (circle), and *L. pudica* Stal (star).

Discussion.—Tower (1906) remarks that this species is closely allied to *L. novemlineata*. Jacoby (1883) gives Oaxaca and Juquila in the state of Oaxaca, Yolotepec, Puebla, and Yolos in the state of Pueble, and Cuernavaca in the state of Morelos, all in Mexico, as localities of *L. dilecta*.

Leptinotarsa obliterata Chevrolat

Doryphora obliterata Chevrolat, 1833. Coleopteres de Mexique 1:23.
Type locality: Orizaba, state of Veracruz, Mexico.

Leptinotarsa subnotata Stal, 1858. Ofv. Svenska Vet.-Akad. Forh. 15:476 (Synonymy, Stal 1863:162.)
Type locality: Mexico.

Chrysomela obliterata Stal, 1863.
Monographie des Chrysomelides de l'Amerique. 2:162 (New generic assignment.)

Leptinotarsa obliterata Jacoby, 1833.
Biologia Centrali-Americana, Insecta, Coleoptera. 6(1):232. (New generic assignment.)

Tower, 1906:8; Blackwelder, 1946:673.

Distinctive features: head dark blue-black, metallic; thorax dark aeneous; head punctation variable, dense; thorax punctation fine, sparse; elytra flavous, each elytron with three dark brown irregular spots; punctation coarse, in double irregular rows; legs and ventral side shinny black, unicolorous.

Head: immaculate, dark blue-black, slightly shinny; punctation fine to coarse, coarse near eyes, dense; interocular distance 2.0 mm, head width 2.4 mm.

Thorax: immaculate, dark aeneous; punctation fine, sparse, anterior-lateral angles blunt or rounded; pronotum length 2.5 mm, pronotum width 5.7 mm. Scutellum dark, aeneous; smooth. Elytra (Fig. 55) oval,

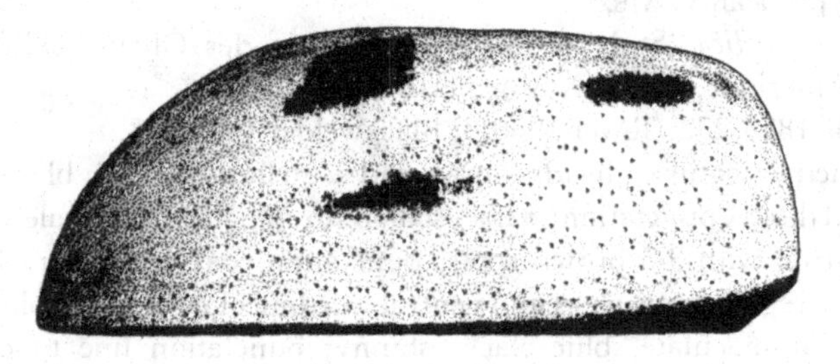

Fig. 55. Left elytron of *L. obliterata* (Chevrolat), dorsal view.

convex; each elytron pale yellow with 3 irregular dark brown spots. One elongated mark located near the humeral angle; other 2 marks at middle of elytron, one a dash like mark nearest the sutural margin and other larger and more rounded; entire margin of elytron outlined in dark brown; upper third of sutural margin near the base has the brown area expanded slightly; punctation in double irregular rows from base to apex, more confused at apex; elytra length 8.4 mm, elytra width 7.8 mm.

Legs: unicolorous, shinny black.

Abdomen: unicolorous, black with white pubescence.

Females.—No external difference.

Individual variation.—Total length 11.5 mm ± .70 (11.0-12.0); greatest width 7.7 mm ± .70 (7.5-8.0); interocular distance 2.0 mm ± .07 (2.0-2.1); head width 2.4 mm ± .14 (2.3-2.5); pronotum length 2.5 mm ± .07 (2.5-2.6); pronotum width 5.7 mm ± .35 (5.5-6.0); elytra length 8.4 mm ± .10 (7.7-9.2); elytra width 7.7 mm ± .35 (7.5-8.0).

The 3 irregular marks on each elytra vary in size from very small marks to a blotch or patch.

Specimen examined.—(Fig. 54, map) 2: MEXICO: 1, (USNM); 1, Fortin de Las Flores, Veracruz, elevation 2500-3000 feet, May 21, 1965 (FSCA).

Discussion.—Jacoby (1883) remarks that this species is exactly like *L. flavitarsis* except that the elytra design is spots instead of vittae. In addition to its resemblance to *L. flavitarsis*, t also resembles *L. pudica* except that *L. pudica* is much smaller.

Tower (1906) reports *L. oblite ata* as characteristic of the lowlands and has studied specimens collected by Salle at Toxpam and Cordoba from Veracruz and one collected by Hoge from Almolonga also in Veracruz.

Leptinotarsa pudica Stal

Leptinotarsa pudica Stal, 1860. Ofv. Svenska Vet.-Akad. Forh. 17:456.
Type locality: Mexico.

Chrysomela pudica (Stal. 1863) Monographie des Chrysomelides de l'Amerique. 2:162 (New generic assignment.)

Jacoby 1883:232; Tower 1906:8; Blackwelder 1946:673.

Distinctive features: head, thorax and scutellum metallic blue-black, head and thorax punctation coarse, scutellum smooth, elytra-pale yellow, each elytron with 3-5 brown spots, punctation coarse and dark brown forming irregular rows, legs and ventral side unicolorous, blue-black.

Head: immaculate, blue-black, shinny; punctation fine to coarse, dense; interocular distance 1.5 mm, head width 1.8 mm.

Thorax: immaculate, blue-black, shinny; punctation coarse, coarser at lateral margins; anterior lateral angles blunt; pronotum length 2.1 mm, pronotum width 4.6 mm. Scutellum blue-black, smooth. Elytra (Fig. 56) oval, convex; each elytron pale yellow with 3-5 irregular brown spots. Two

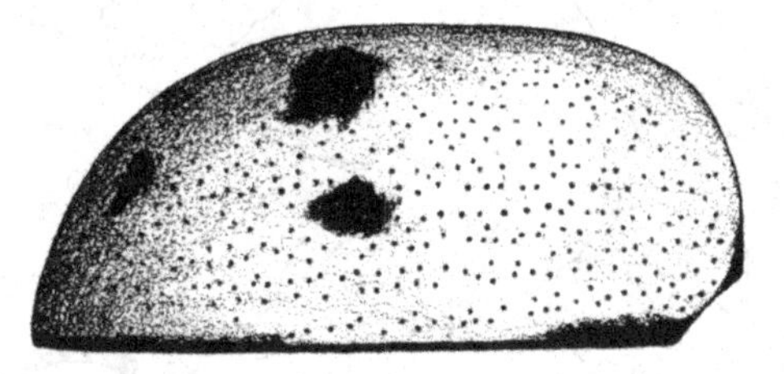

Fig. 56. Left elytron of *L. pudica* Stal, dorsal view.

spots located at the disk and one at apex of elytron; margin of elytra black including sutural margin; punctation dense, forming irregular rows, punctation darkened; elytra length 7.2 mm, elytra width 6.4 mm.

Legs: unicolorous, shinny, blue-black.

Individual variation.—Only one specimen examined; total length 9.2 mm, greatest width 6.4 mm, interocular distance 1.5 mm, head width 1.8 mm, pronotum length 2.1 mm, pronotum width 4.6 mm, elytra width 6.4 mm, elytra length 7.2 mm.

Specimens examined.—(Fig. 54, map) 1: MEXICO: 1, Juquila, (USNM).

Discussion.—This species is distinguished from *L. obliterata* by the smaller size and coarser, more irregular and darkened punctation. Jacoby (1883) studied a specimen with 5 spots collected by Hoge at Cordova. The specimen I have studied was collected by Salle at Juquila and has only 3 distinct spots, none at the base of the elytra. This specimen is from the Monros collection.

Leptinotarsa dohrni Jacoby

Leptinotarsa dohrni Jacoby, 1883. Biologia Centrali-Americana, Insecta, Coleoptera, Chrysomelidae 6(1):239.

Type locality: Yolotepec, Mexico. (Fig. 57, map.)

Tower, 1906:4; Blackwelder, 1946:673.

Distinctive features: head and pronotum cupreous, sparsely punctured, elytra testaceous, covered with small and large black spots, ventral side metallic green, antennae, apex of tibia and tarsi flavous.

Head: unicolorous, cupreous; punctation fine; labrum, maxillary palps, antennae flavous.

Thorax: unicolorous, cupreous; punctation sparse; anterior-lateral

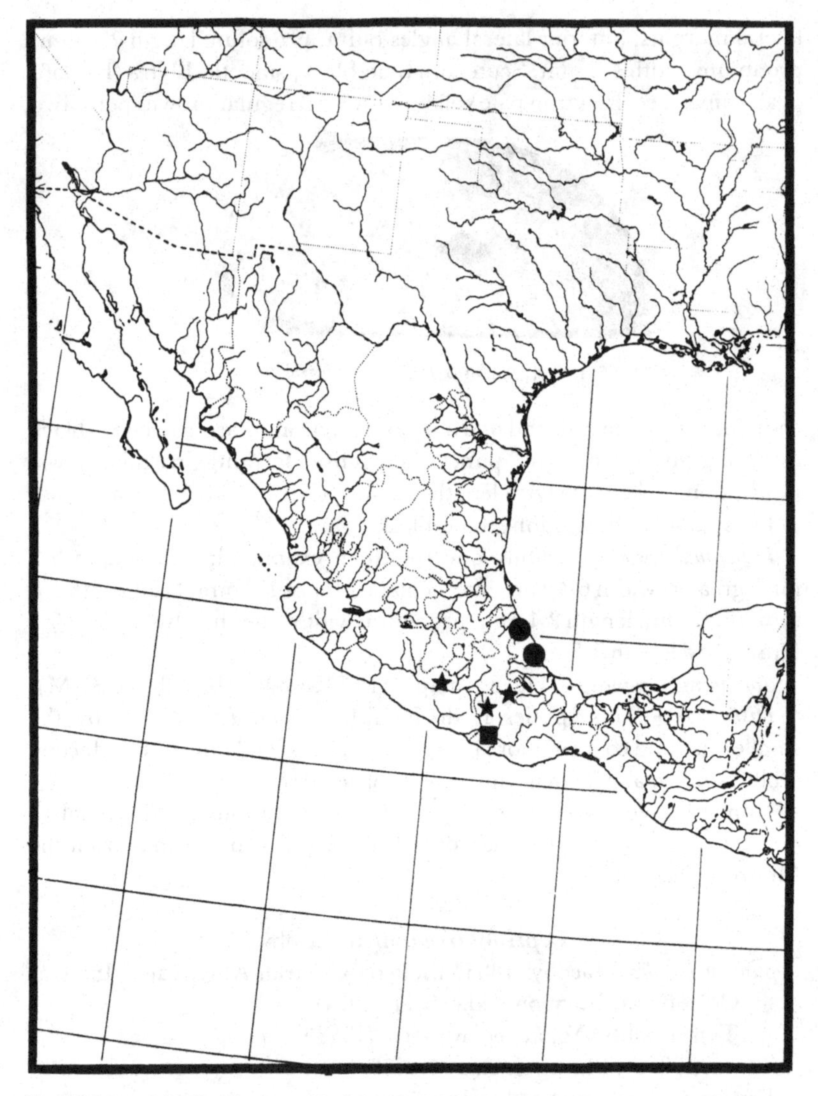

Fig. 57. Map of the distribution of *L. dohrini* Jacoby (square), *L. hogei* Jacoby (circle), and *L. stali* Jacoby (star).

angles rounded. Scutellum cupreous, smooth. Elytra oval, very convex; base color of elytra testaceous; each elytra with 10 rows of small black spots placed near sutural and lateral margins, larger black confluent spots placed at middle; geminately punctate-striate.

Legs: mulicolorous, basal parts metalic green to bronze; apex of tibia and tarsi flavous.

Abdomen: metallic green to bronze.

Individual variation.—No specimens were available for study. Jacoby recorded this species as 4.5 lines (= 9.0 mm) in length.

Specimens examined.—None, description from original description.

Discussion.—Jacoby (1883) described this species from a single specimen collected by Salle. He noted that the curious markings on the elytra are subject to variation. Tower (1906) records this species as rare, and also studied only 1 specimen, the same one that Jacoby studied and described. Tower questions the locality of this species. Salle only placed Yoltepec on the label, but there are two Yoltepecs, one in the state of Oaxaca and one in the state of Hidalgo.

Leptinotarsa hogei Jacoby

Leptinotarsa hogei Jacoby, 1883. Biologia Centrali-Americana, Insecta, Coleoptera, Chrysomelidae. 6(1):240.

Type locality: Cerro de Plumas, Mexico.

Tower, 1906:4; Blackwelder, 1946:673.

Distinctive features: head and thorax aeneous; elytra aeneous, each elytron with 2 large flavous bands extending from base to apex; elytra punctation coarse in aeneous areas only; legs aeneous, tarsal segments flavous; ventral surface aeneous.

Head: immaculate, aeneous, shinny; punctation fine, dense; interocular distance 1.7 mm, head width 2.2 mm.

Thorax: immaculate, aeneous, punctation coarse, coarser and denser at lateral margins; anterior-lateral angles pointed; pronotum length 2.0 mm, pronotum width 4.8 mm. Scutellum aeneous, outlined in metallic green, smooth. Elytra (Fig. 58) outlined in aeneous, wider along sutural

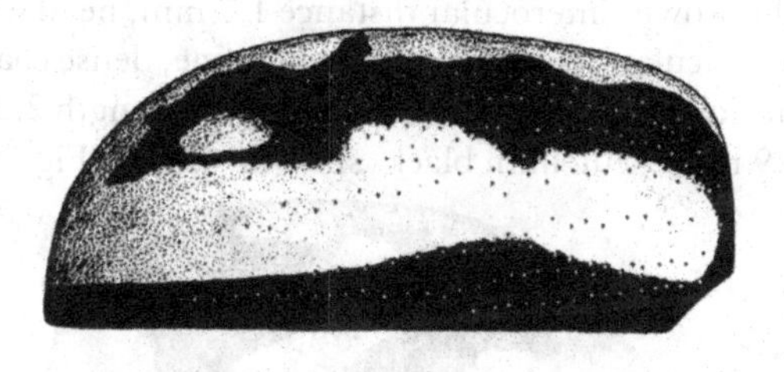

Fig. 58. Left elytron of *L. hogei* Jacoby, dorsal view.

margin, two flavous bands, 1 mm in width extend from base of elytron and both join at apex; aeneous band in between flavous band with course punctation in irregular rows, basal end of aeneous band with flavous spot, margins of this band irregular; no punctation in the flavous areas; elytra length 7.4 mm, width 6.5 mm.

Legs: aeneous, tarsal segments flavous.

Abdomen: unicolorous, aeneous with white pubescence.

Individual variation.—Only one specimen examined; total length 9.6 mm, greatest width 6.5 mm, interocular distance 1.7 mm, head width 2.2 mm, pronotum length 2.0 mm, pronotum width 4.8 mm, elytra length 7.4 mm, elytra width 6.5 mm.

Specimen examined.—(Fig. 57, map) 1: MEXICO: 1, Veracruz, Cerro de Plumas, (USNM).

Discussion.—Jacoby (1883) and Tower (1906) both remark that this species bears a close resemblance to *Zygogramma ornata* Jacoby (1882) but is clearly in the genus *Leptinotarsa*. Jacoby (1883) described this species from 10 specimens obtained from Hoge. This species is known from only one locality, Cerro de Plumas, Mexico.

Leptinotarsa stali Jacoby

Leptinotarsa stali Jacoby, 1883. Biologia Centrali-Americana, Insecta, Coleoptera, Chrysomelidae. 6(1):237.

Type locality: Izucar, State of Puebla, Mexico.

Tower, 1906:4; Blackwelder, 1946:673.

Distinctive features: head and thorax aeneous; elytra black with yellow-orange markings, each elytron with a large orange-yellow spot at base of elytra and two yellow-orange transverse bands at disk and base of elytron; apical margin of elytra orange-yellow, no orange-yellow along sutural margin; legs and ventral surface unicolorous.

Head: immaculate, aeneous; punctation fine, dense; clypeus golden; antennae light brown; interocular distance 1.7 mm, head width 2.2 mm.

Thorax: immaculate, aeneous; punctation fine, dense coarser at lateral margins; anterior-lateral angles blunt; pronotum length 2.3 mm, pronotum width 4.9 mm. Scutellum black, smooth. Elytra (Fig. 59) black with

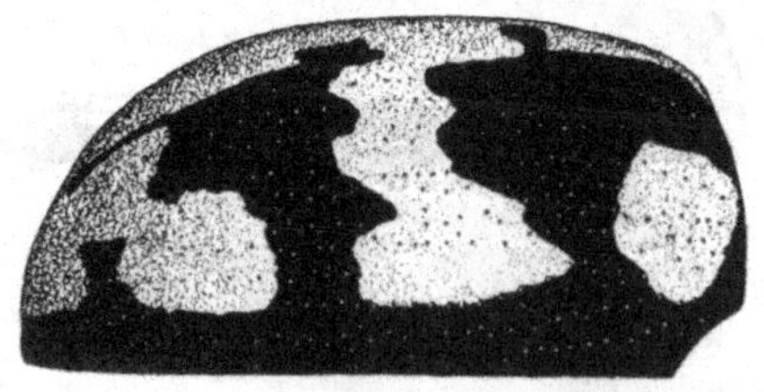

Fig. 59. Left elytron of *L. stali* Jacoby, dorsal view.

orange-yellow markings, each elytra with a large orange-yellow spot at base of elytra; lateral margins of elytra orange-yellow transverse bands extend from lateral margins to sutural margin but not reaching it, no orange-yellow on the sutural margin. Punctation coarse and forming rows in the black area only, very faint punctation and sparse in the orange-yellow areas; elytra length 7.0 mm, width 6.5 mm.

Legs: unicolorous, aeneous.

Abdomen: unicolorous, black.

Individual variation.—Only one specimen examined, total length 8.5 mm, greatest width 6.5 mm, interocular distance 1.7 mm, head width 2.2 mm, pronotum length 2.3 mm, pronotum width 4.9 mm, elytra length 7.0 mm, elytra width 6.5 mm.

Specimens examined.—(Fig. 57, map) 1: MEXICO: 1, (USNM).

Discussion.—Jacoby (1883) described this species from two specimens in the Salle collection. Tower (1906) remarks that this species is known only from the head waters of the Rio Atoyac and the Rio Coetzala. He collected specimens from Pueble, Matamorus de Izuar and Atlizco all in the state of Pueble.

Leptinotarsa lineolata (Stal)

Chrysomela lineolata Stal. 1863. Monographie des Chrysomelides de l'Amerique. 2:159.

Type locality: Texas.

Myocoryna lineolata Crotch, 1873. Proc. Acad. Nat. Sci. Philadelphia. 25:47. (New generic assignment.)

Leptinotarsa lineolata Jacoby, 1891. Biologia Centrali-Americana, Insecta, Coleoptera. 6(1):253. (New generic assignment.)

Linell, 1896:196; Tower, 1906:8; Leng, 1920:294; Blackwelder, 1946:673; Wilcox 1972:9.

Distinctive features: head, thorax, and scutellum metallic green to aeneous; head and thorax densely punctured; elytra pale yellow, each with 4 brown to aeneous interrupted vittae; abdomen aenous; legs light brown (Fig. 60).

Head: unicolorous, metallic green to aeneous; antennae light brown, distal 5 segments darkened, expanded; punctation coarse, dense; interocular distance 1.4 mm, head width 1.7 mm.

Thorax: unicolorous, metallic green to aeneous; punctation fine at disk, coarser at lateral margins, dense; anterior-lateral angle pointed slightly or blunt; pronotum length 1.7 mm, pronotum width 3.9 mm. Scutellum metallic green to aeneous, usually smooth but a few punctures may be

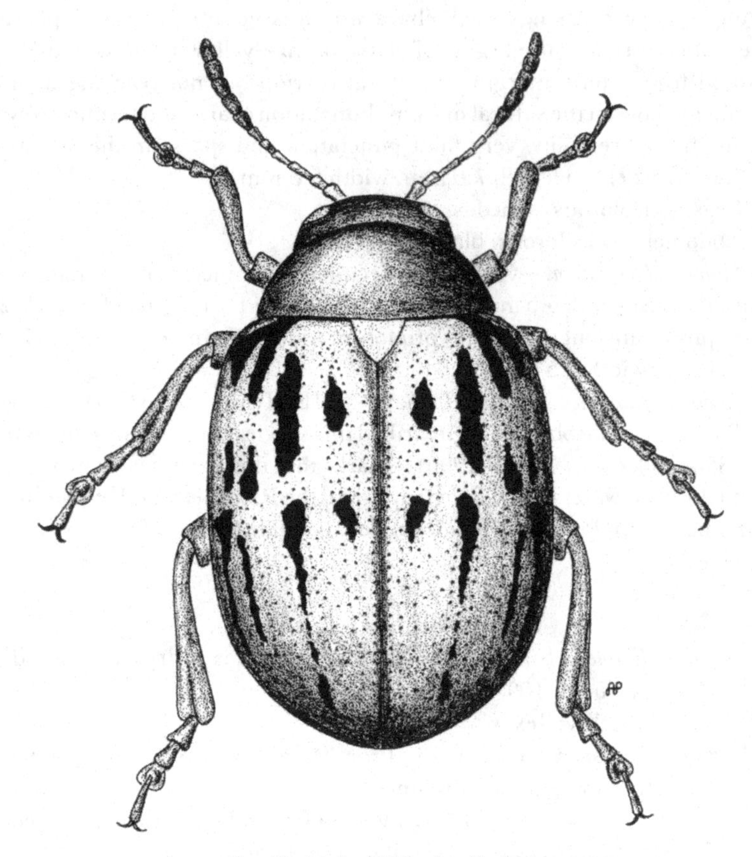

Fig. 60. Adult of *L. lineolata* (Stal), dorsal view.

present. Elytra pale yellow; punctation coarse, not forming rows; each elytron with 4 brown to aeneous interrupted vittae, second vitta most prominent and longest, thickened at base of elytron, third and fourth vittae join at base of elytron; elytra length 6.0 mm, elytra width 5.4 mm.

Legs: unicolorous, light brown.

Abdomen: unicolorous, aeneous.

Male genitalia.—Aedeagus a flattened cylinder, gently arched; apical half greatly flattened and expanded, apex rounded (Figures 61 and 62).

Females.—Generally larger than males.

Individual variation.—Total length 7.9 mm ± .42 (7.0-8.6); greatest width 5.4 mm ± .36 (4.7-6.0); interocular distance 1.7 mm ± .07 (1.3-

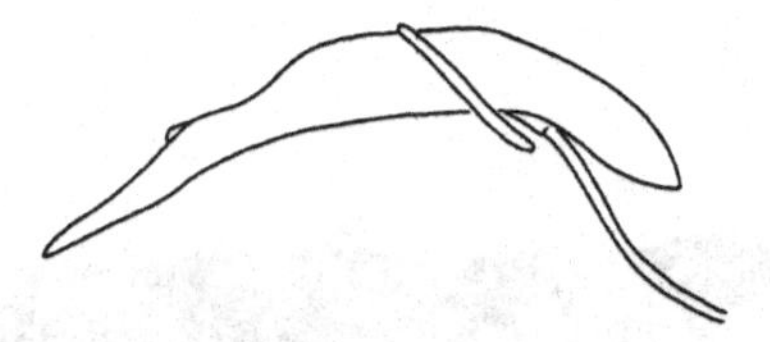

Fig. 61. Male genitalia of *L. lineolata* (Stal), lateral view.

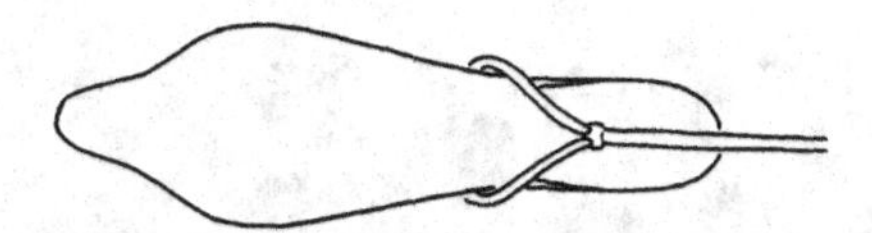

Fig. 62. Male genitalia of *L. lineolata* (Stal), ventral view.

1.5); head width 1.7 mm ± .08 (1.6-1.9); pronotum length 1.7 mm ± .15 (1.5-2.0); pronotum width 3.9 mm ± .19 (3.6-4.2); elytra length 6.0 mm ± .39 (5.2-6.9) pronotum width 5.4 mm ± .36 (4.7-6.0).

Color of the head, thorax and scutellum varies from green to a bronze color. Vittae on the elytra without much variation.

Biology.—During the summer of 1971 studies on the biology of this species were conducted in an area near Nogales, Arizona. Previous studies on this species were conducted by R. H. Arnett, Jr. and E. R. Van Tassell during the summers of 1961-1964 in the Peña Blanca area just north west of Nogales. These field studies, supported by rearing data, piece together the life cycle.

Leptinotarsa lineolata feeds exclusively on *Hymenoclea monogyra* (Compositae) (Fig. 63), a common plant restricted to sandy washes in the southwest United States and the northern Mexican Basin.

The area of study is located 1 mile north of Nogales, Arizona in Mariposa canyon. Species of *L. lineolata* were found feeding on *Hymenoclea monogyra* Torrey and Gray throughout the canyon. *H. monogyra* is found in sandy soil from western Texas to southern California and Northern Mexico at elevations from 1000 to 4000 feet and is commonly called burro-brush. Burro-brush is a low much branched shrub and tends to form thickets. Thickets average six foot square. The plants ranged from 24 to 60 inches high in Mariposa canyon. Since *L. lineolata* feeds exclusively on the burro-brush, it is here referred to as the burro-brush beetle.

Female burro-brush beetles lay their eggs in tightly glued clusters on the upper parts of the tubular burro-brush leaves (Fig. 64). Eggs hatch in an

Fig. 63. A stand of *Hymenoclea monogyra* Torre & Gray, near Nogales, Arizona, a breeding site of *L. lineolata* (Stal).

Fig. 64. Adult of *Perillus bioculatus* (Fab.) (Hempitera: Pentatomidae), feeding on the eggs of *L. lineolata* (Stal).

average of 7 days in the field, first instar larvae remain close to the site of hatching for at least 24 to 48 hours. There are 4 instars.

The larvae feed on the leaves (Fig. 65) and move rapidly along the upper

Fig. 65. Larva of *L. lineolata* (Stal) feeding on the leaves of *Hymenoclea monogyra* Torre & Gray.

parts of the plant. They are easy to detect by their black stripes. It takes approximately 15 days from hatching until the fourth instar drops to the ground and burrows into the soil. Observations of the fourth instar show a transformation into a prepupa stage before final pupation. The prepupa are not active, slightly curved, and look much like the fourth instar larvae except it is buried in the soil. The pupa (Fig. 66) is milky white. About 10

Fig. 66. Pupa of *L. lineolata* (Stal), dorsal view.

days are required for transformation. The entire life cycle from the time eggs are laid to the adult emergence is approximately 32 days. Adults (Fig. 60) are found on all parts of the plant above the soil. When disturbed they drop to the soil, a common occurrence of many chrysomelid beetles.

Specimens examined.—(Fig. 67, map) 963: ARIZONA: Santa Cruz Co.

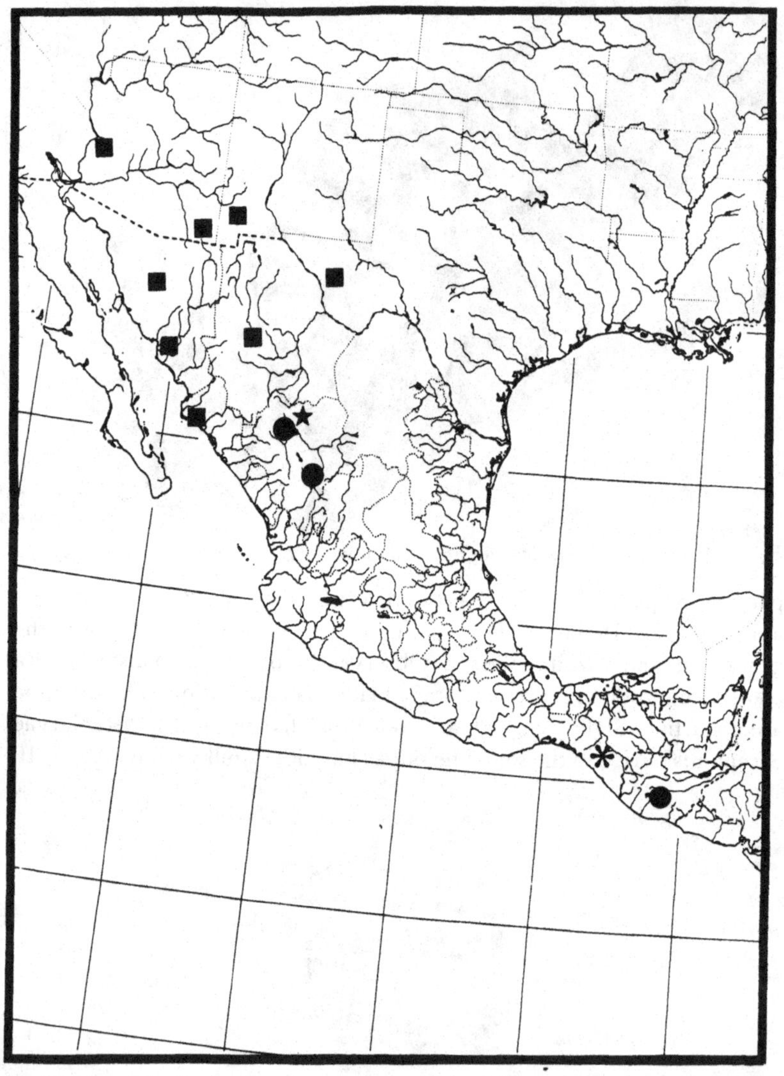

Fig. 67. Map of the distribution of *L. lineolata* (Stal) (square), *L flavitarsis* (Guer.-Meneville) (circle), *L. similis* Bowditch (star), and *L. virgulata* Achard (asterisk).

56, 1 mi. N. Nogales, Aug. 1-24, 1971 (RLJC); 72, Peña Blanca, Pajarito Mts., Aug. 17, 1971 (RLJC); 72, Nogales, Aug. 12, 1900 (USNM); 5, Nogales, Aug. 29, 1965 (USNM); 10, Arivaca, July 27, 1961 (USNM); 5, 7 mil N. Nogales, Sept 12, 1958 (CISC); 8, 10 mi. SW. Patagonia, Sept. 13, 1958 (CISC); 1, Patagonia, Sept. 21, 1961 (CISC); 93, Peña Blanca, Pajarito Mts. July 25, 1961 (FSCA); 5, Nogales, Aug. 2, 1955 (OSUC); 2, Patagonia July 8, 1949 (OSUC); 6, 10 mi. E. Nogales, Sept. 2, 1957 (UDCD); 1, Patagonia, Aug. 30, 1954 (UCDC); 3, 15 mi. NW Nogales, Sept. 8, 1957 (UCDC); 42, 10 mi. E. Nogales, July 28, 1956 (AMNH); 3, 5 mi. NE. Nogales, Aug. 25, 1950 (AMNH; 2, Nogales, Aug. 10, 1913 (CASC); 1, Pena Blanca, Sept. 6, 1968 (USUC); Cochise Co. 16, Douglas, Aug. 8, 1932 (CISC); 20, Walnut Gulch, Tombstone, July 17, 1950 (CISC); 6, 8 mi. E. Douglas, Aug. 4, 1958 (CISC); 1, Portal, Sept. 1, 1961 (CISC); 5, Dos Cabezas, July 31, 1952 (AMNH); 2, 5 mi. SE. Apache, Aug. 11, 1958 (UCDC); Pima Co. 18, Oracle, July 13, 1954 (CISC); 11, Box Canyon, July 30, 1964 (CISC); Yavapai Co. 2, Cottonwood, Aug. 30, 1961 (CISC); 1, 10 mi. NW. Congress, May 30, 1962 (UDCD); Pinal Co. 2, 14 mi. E. Oracle, July 27, 1924 (CASC); NEW MEXICO: 2, 15 Mi. S. Animas, Hildalgo Co. July 31, 1965 (CASC); TEXAS: Davis Co. 16, Davis Mts., June 20-24, 1949 (OSUC); 1, White Rose Canyon, June 18, 1947 (UCDC); MEXICO: CHIHUAHUA: 90, 25 mi. SW. Camargo, July 14, 1947 (AMNH); 109, 20 mi. SW. Camargo, July 13, 1947 (AMNH); 51, Delicias, July 11, 1947 (AMNH); 20, LaCruz, July 13, 1947 (AMNH); SONORA: 11, Montezuma, June 4, 1954 (CISC); 4, 20 mi. E. Hermosilla, July 20, 1952 (AMNH); 2, SE. Alamox, Sept. 5, 1970 (UDCD); 84, N. Novojos, Aug. 2, 1952 (AMNH); 86, Navojoa, Aug. 3, 1952 (AMNH); SINALOA: 13, Elota, June 27, 1962 (CISC); 1, Culiacan, July 17, 1962 (CISC); 2, 76 mi. S. Culican, July 25, 1952 (CISC).

Discussion.—The range of this species is limited to the Northern Mexican Basin, and the southern sections of Arizona and New Mexico and western sections of Texas.

Jacoby (1891) reported this species in Texas, Sonora, and Chihuahua. Tower (1906) gives a similar range, and Bristley (1926) reported *L. lineolata* common on *Hymenoclea* sp. in Arizona.

L. lineolata resembles *L. flavitarsis* in color and elytra markings but it is much smaller.

Leptinotarsa flavitarsis (Guerin-Meneville)

Polyspila flavitarsis Guerin-Meneville, 1855. Verh. Zool.-Bot. Ges. Wien. 5:606.

Type locality: Mexico.

Leptinotarsa signatipennis Baly, 1858. Trans. Ent. Soc. London. 4(2):352. (Synonymy, Stal 1863:161.)

Type locality: Mexico.

Leptinotarsa flavitarsis Stal, 1858. Ofv. Svenska Vet.-Akad. Forh. 15:476. (New generic assignment.)

Leptinotarsa nitidicollis Stal, 1860. Ofv. Svenska Vet.-Akad. Forh. 17:456. (NEW SYNONYMY.)

Type locality: Mexico.

Stal, 1863:160; Jacoby, 1883:238: Tower, 1906:8; Blackwelder, 1946:673.

The shinny metallic blue or green pronotum and head, variable punctation, pale yellow elytra with unique vittae of variable thickness, coarse punctation in irregular rows, legs with proximal half metallic blue or green and distal half yellow will separate this species from all others in North America.

Head: punctation variable; proximal 6 segments of antennae slender, yellow, distal 5 segments expanded, dull covered with fine pubescence, interocular distance 1.8 mm, head width 2.0 mm; eyes feebly emarginate.

Thorax: immaculate, shinny, smooth, blue or green. Elytra (Fig. 68)

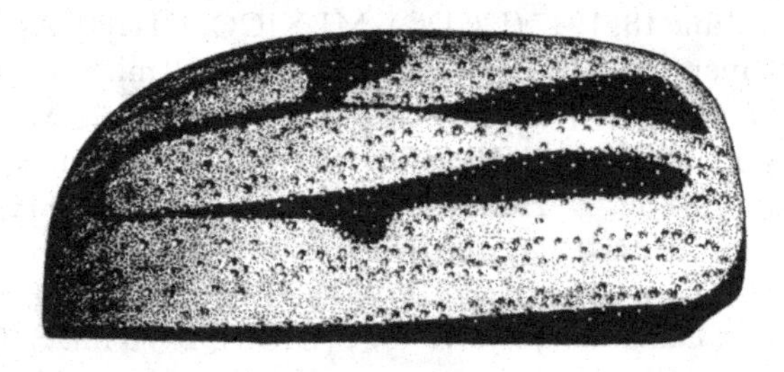

Fig. 68. Left elytron of *L. flavitarsus* (Guer.-Meneville), dorsal view.

oval, convex, pale yellow; elytra margins black, 2 vittae extend from base of the elytron joining at apex, a third vitta near side of disk is one half base of elytron; vitta 1 with thickened area midway, extending towards sutural margin; vitta 3 thickened at base and sometimes joins with vitta 2 at its midpoint; vittae dark brown to black.

Legs: coxae, femur, and proximal half of tibia metallic blue or green; distal half of tibia and tarsus golden.

Abdomen: metallic blue or green.

Individual variation.—Total length 10.3 mm ± .40 (9.7-11.0); greatest width 7.1 mm ± .39 (6.5-7.5); interocular distance 1.8 mm ± .10 (1.7-

2.0); head width 2.0 mm ± .08 (1.9-2.2); pronotum length 2.1 mm ± .14 (2.0-2.2); pronotum width 5.1 mm ± .22 (4.7-5.4); elytra length 8.1 mm ± .35 (7.5-8.6); elytra width 7.1 mm ± .39 (6.5-7.5).

Specimens examined.—(Fig. 67, map) 9: MEXICO: 2 (USNM); GUATEMALA: 5 (USNM); CENTRAL AMERICA: 2 (USNM).

Discussion.—Baly (1858) names Mexico as the type locality but Jacoby (1883) casts doubt on this species being in Mexico at all by remarking that all the specimens he ever examined were collected in Central America. Tower (1906) suggested that *L. nitidicollis* was a variety of *L. flavitarsis*. *L. nitidicollis* is placed in synonymy with *L. flavitarsis* after careful examination of specimens.

Leptinotarsa similis Bowditch

Leptinotarsa similis Bowditch, 1911. Trans. American Ent. Soc. 37:326.
Type locality: Mexico.
Blackwelder, 1946:673.

Distinctive features: head and thorax metallic green, punctation dense; elytra brown with metallic green maculation, each elytron with complete vittae near sutural margin and 2 abbreviated vittae at base of elytron; each elytron with spots at disk and apex of elytron; legs golden brown with a large green patch on femur, ventral surface dark green.

Head: immaculate, metallic green; punctation sparse to dense, coarser near eyes; antennae, labrum and clypeus flavous; interocular distance 1.4 mm, head width 1.7 mm.

Thorax: immaculate, metallic green, punctation dense, fine, coarser near lateral margins; anterior-lateral angles rounded; pronotum length 1.8 mm, pronotum width 4.2 mm. Scutellum metallic green, smooth. Elytra (Fig. 69) oval, convex; base color of elytra light brown with metallic

Fig. 69. Left elytron of *L. similis* Bowditch, dorsal view.

green maculation, each elytron with a metallic green area along base of sutural margin, one nearly complete vittae at base of elytron, second with a hooked apex, disk and apex of elytron with at least 20 spots; humeral angles with a green area; punctation coarse, very irregular; elytra length

6.5 mm, elytra width 5.8 mm.

Legs: flavous with a large green patch on femur.

Abdomen: unicolorous, dark aeneous.

Individual variation.—Only one specimen examined; total length 8.5 mm, greatest width 5.8 mm, interocular distance 1.4 mm, head width 1.7 mm, pronotum length 1.8 mm, pronotum width 4.2 mm, elytra length 6.5 mm, elytra width 5.8 mm.

Specimens examined.—(Fig. 67, map) 1: MEXICO: 1, Sept. 16, 1985 (USNM).

Discussion.—Bowditch (1911) described this rare species from Jacoby material. The markings of this species are unique, although *L. typographica* has similar markings. This species may resemble certain color forms of *L. dilecta* except that *L. dilecta* lacks spots on the elytra.

Leptinotarsa virgulata Achard

Leptinotarsa virgulata Achard, 1923. Nouveaux Chrysomelini d'Amerique. Fragments entomologiques. 5:67.

Type locality: State of Chiapas, Mexico (Fig. 67, map.)

Blackwelder, 1946:673.

Distinctive features: head, pronotum, and femur of legs metallic green or coppery-green; palps, base of antennae, part of tibia and part of metasternum flavous; elytra yellow-brown with a narrow sutural band and 2 fine vittae; head and pronotum strongly punctured; scutellum finely spotted; abdomen yellow-brown with fine punctation.

Head: unicolorous, metallic green, having coppery or bronze appearance; punctation coarse and dense; maxillary palps and distal part of antennae yellow-brown.

Thorax: unicolorous, metallic green, having coppery or bronze appearance; punctation coarse and dense; anterior-lateral angles intense, dent or depression near center of pronotum. Scutellum finely spotted or stippled. Elytra oval, convex; base color brownish-yellow; each elytron with a narrow sutural band and two fine vittae, outer vittae shorter than inner and joining oblique mark at elytra apex, an elongated comma mark near center of elytra; elytra punctation coarse, finer along sides.

Legs: middle of femur metallic green; part of tibia, except the base, yellow-brown.

Abdomen: yellow-brown; punctation fine.

Individual variation.—No specimens were available for study. Achard records the total length as from 9.0-11.0 mm.

Specimens examined.—None, description from original description.

Discussion.—This species is apparently rare. Achard says this species is similar to *L. calceata* and *L. novemlineata*, both described by Stal.

LITERATURE CITED

Achard, J. 1923. Nouveaux Chrysomelini d'Amerique. Fragments entomologiques. 5:65-80.

Arnett, R. H. Jr. 1947. A technique for staining, dissecting, and mounting the male genitalia of beetles. Col. Bull. 1(7):63-66.

Arnett, R. H. Jr. 1963. The beetles of the United States (A manual for identification). The Catholic Univ. of America Press. Washington, D. C. 1112 pp.

Arnett, R. H. Jr. 1985. American Insects, A Handbook of the Insects of America North of Mexico. Van Nostrand Reinhold Company, New York, N. Y. 850 p.

Arnett, R. H. Jr., Downie, N. M., and H. E. Jaques. 1980. How to Know the Beetles, 2nd ed., Wm. C. Brown Co., Dubuque, Iowa viii + 416 p.

Arnett, R. H. Jr., and G. A. Samuelson (ed.). 1986. The Insect and Spider Collections of the World. E. J. Brill/Flora & Fauna Publications, Gainesville, Florida. 220 pp.

Bailey, T. and Kok, L. 1978. Insects Associated with Horse Nettle, *Solanum carolinense*, in Southwest Virginia. 56th Annual Meeting of the Virginia Academy of Science, Virginia Journal of Science, 29 (2).

Baly, J. S. 1858. Descriptions of some new species of Chrysomelidae. Trans. Ent. Soc. London 4(2):293-352.

Barber, H. S. and Bridwell, J. C. 1940. Dejean catalogue names (Coleoptera). Bull. Brooklyn Ent. Soc., 35(1):1-12.

Barber, G. W. 1933. Insects attacking *Solanum sisymbriifolium* in eastern Georgia. J. Econ. Ent. 26:1174-1175.

Bernon, Gary. 1985, 1987 (Personal Communication).

Blackwelder, R. E. 1946. Checklist of the coleopterous insects of Mexico, Central America, the West Indies, and South America, Part 4. Bull. United States Nat. Mus., 185:551-763.

Blackwelder, R. E. 1957. Checklist of the coleopterous insects of Mexico, Central America, the West Indies, and South America. Part 6. Bull. United States Nat. Mus., 185:927-1492.

Blatchley, W. S. 1910. The Coleoptera or beetles of Indiana. Bull. Indiana Dept. Geol. Nat. Res., 1386 p.

Borror, D. J. and White, R. E. 1970. A field guide to the insects of America north of Mexico. Houghton Mifflin Co., Boston. 404 pp.

Bowditch, F. C. 1911. Notes on *Callilgrapha* and its allies. Trans. American Ent. Soc. 37:325-334.

Bradley, J. C. 1930. A manual of the genera of beetles of America north of Mexico. Ithaca, New York. 360 pp.

Brisley, H. R. 1926. Notes on the Chrysomelidae (Coleoptera) of America. Trans. American Ent. Soc. 51:167-182.

Brown, W. J. 1945. Food plants and distribution of the species of *Calligrapha* in Canada, with descriptions of new species. Canadian Ent., 77:117-133.

Brown, W. J. 1961. Notes on North American Chrysomelidae. Canadian Ent., 93:967-977.

Brown, W. J. 1962. The American species of *Chrysolina* Mots. (Coleoptera:Chrysomelidae). Canadian Ent., 94:58-74.

Brues, C. T. 1940. Food preferences of the Colorado potato beetle, *Leptinotarsa decemlineata* (Say). Psyche 47:38-43.

Cass, L. M. 1957. Dispersal of larvae of the Colorado potato beetle, *Leptinotarsa decemlineata* (Say). Canadian Ent., 89:581-582.

Chevrolat, L. A. A. 1833. Coleopteres du Mexique, fasc. 1, Strasbourg. 25 pp.

Chevrolat, L. A. A. 1877. Descriptions de novelles especes de coleopteres, Ann. Soc. Ent. France. 7(5):97-104.

Chittenden, F. A. 1924. The return of *Leptinotarsa juncta* (Germar) to the District of Columbia. Bull. Brooklyn Ent. Soc. 19:37.

Correll, D. S. and Johnston, M. C. 1970. Manual of the Vascular Plants of Texas., Texas Research Foundation, Renner, Texas. 1881 p.

Crotch, G. R. and Cantab. M. A. 1873. Materials for the study of the Phytophaga of the United States. Proc. Acad. Nat. Sci. Philadelphia. 25:19-83.

Dejean, P. F. M. A. 1836. Catalogue des coleopteres de la collection de M. le comte Dejean. Livre 4:257-360.

Drummond, F., Casagrande, R., Chauvin, R., Hsiao, T., Lashomb, J., Logan, P., and Atkinson, T. 1984. Distribution and New Host (Acari: Tarsonemina; Podapolipidae) Attacking the Colorado Potato Beetle in Mexico. Internat. J. Acarol. 10:3, 179-180.

Fitch, A. 1863. Ninth Report on the Noxious and other Insects of the State of New York. Transactions of the New York State Agricultural Society 5:778-823.

Gauthier, N. L., Hoffmaster, R., and Semel, M. 1981. History of Colorado potato beetle control. In Advances in potato pest management. J. H. Lashomb and R. Casagrande (eds.) Hutchinson Ross Publ. Co., Stroudsburg, Pa., pp. 13-33.

Germar, E. F. 1824. Insectorum species novae ant minus cognitae, 624 pp.

Gidaspow, T. 1959. North American caterpillar hunters of the genera *Calosoma* and *Callisthenes* (Coleoptera: Carabidae). Bull. American Mus. Nat. Hist., 116(3): 225-344.

Guerin-Meneville, F. E. 1855. Catalogue des insects coleopteres. Verhandlungen der k.k. zoologisch-botanischen Gesellschaft, Wien. 5:573-612.

Harold, E. von. 1877. Coleopterorum species novae. Mitth. Munchener Ent. Ver. 1:16.

Heiser, C. 1969. Nightshades: the paradoxical plants. W. H. Freeman and Co., San Francisco. 200 p.

Horn, G. H. 1894. The Coleoptera of California. Proc. California Acad. Sci. 4(2):302-449. Hsiao, T. H. 1974. Chemical influences of feeding behavior *Leptinotarsa* beetles. In Experimental Analysis of Insect Behavior, L. B. Browne, ed. pp. 237-248. New York: Springer-Verlag.

Hsiao, T. 1981. Ecophysiological adaptations among geographic populations of the Colorado potato beetle in North America. In Advances in potato pest management. J. H. Lashomb and R. Casagrande (eds.). Hutchinson Ross Publ. Co., Stroudsburg, Pa., pp. 69-85.

Hsiao, T. H. and Hsiao, C. 1983. Chromosomal analysis of *Leptinotarsa* and *Labidomera* species (Coleoptera: Chrysomelidae). Genetica 60: 139-150.

Hsiao, T. 1986. Specificity of Certain Chrysomelid Beetles for Solanaceae. In Solanaceae: Biology and Systematics; Second International Symposium. Columbia University Press, New York 345-363.

Illiger, J. C. W. 1807. Vorschlag zur Aufnahme in Fabricischen Systeme Fehlender Kafergattungen. Mag. Insektenk., 6:331.

International Code of Zoological Nomenclature, 3rd ed. adopted by the General Assembly of the International Union of Biological Sciences. International Trust for Zoological Nomenclature, London. 1985. 388 pp.

Isely, D. 1960. Weed Identification and Control in North Central States. Iowa State University Press, Ames, Iowa, 2nd ed., 400 p.

Jacoby, M. 1877. Descriptions of new species of phytophagous Coleoptera. Proc. Zool. Soc. London. 510-520.

Jacoby, M. 1879. Descriptions of new species of phytophagous Coleoptera. Proc. Zool. Soc. London. 773-793.

Jacoby, M. 1883. Biologia Centrali-Americana, Insecta. Coleoptera, Chrysomelidae. vol. 6, part 1:225-264.

Jacoby, M. 1891. Biologia Centrali-Americana, Insecta, Coleoptera, Chrysomelidae, Supplement to Phytophaga. vol. 6, part 1:233-312.

Jacoby, M. 1903. Descriptions of new species of South American Coleoptera of the family Chrysomelidae. Proc. Zool. Soc. London. 2:30-59.

Jacques, R. L. 1985. The Potato Beetles of Florida (Coleoptera: Chrysomelidae) Fla. Dept. Agric. & Consumer Serv. Division of Plant Industry, Entomology Circular No. 271. 2 p.

Jolivet, P. 1986. Insects and Plants: Parallel Evolution and Adaptations. Flora and Fauna Handbook No. 2., E. J. Brill/Flora and Fauna Publications, New York New York. 197 pp.

Jolivet, P. and Petitpierre E. 1976. Selection trophique et evolution chromosomique chez les Chrysomelinae (Col. Chrysomelidae). Acta Zool. Pathol. Antverp. 66:59-60.

Kearney, T. H. and Peebles, R. H. 1960. Arizona Flora. Univ. of California Press. Berkeley and Los Angeles. 1085 pp.

Knab, F. 1907. Notes on *Leptinotarsa undecemlineata* (Stal). J. New York Ent. Soc. 15:190-193.

Knab, F. 1908. Tower's Evolution in *Leptinotarsa*. Science, (n-s), Lancaster, Pennsylvania. 27:223-227.

Kraatz, G. 1874. Zur Nomenclatur des Kartoffelkafers. Berliner Ent. Zeitschr. 18:442-444.

Latreille, P. A. 1801. Considerations generales su l'ordre naturel del animaux composant les classes des crustaces, des archnides, et des insects; avec un tablueau methodique de leurs genres disposes en families. 444 pp.

Leech, H. R. and Green. J. W. 1955. Plant Association Data for a new Arizona and New Mexico Coleoptera. Coleopt. Bull. 9(2): 27-28.

Leng, C. W. 1920. Catalogue of Coleoptera of America, north of Mexico. Mount Vernon, New York. 470 pp.

Linell, M. L. 1896. A short review of the Chrysomelas of North America. J. New York Ent. Soc. 4:195:200.

Logan, P. 1983. Notes from field studies in Mexico, August 17-Sept. 15, 1983. Collaborators include: R. Casagrande, T. Hsiao, T. Martin, R. Chauvin, J. Lashomb, and F. Drummond.

Metcalf, C. L. and Flint W. P. 1939. Destructive and Useful Insects, their habits and control. McGraw-Hill Book Co. Inc. New York. 981 pp.

Monros, F. 1955. On some new genera of Nearctic Chrysomelinae (Chrysomelidae). Coleopt. Bull. 9(4):53-60.

Motschulsky, V. 1860. Coleopteres de la Siberie orientale et particulier des rives de l'Amour. In Schrenck, Reisen and Forschungen im Amurlande. 2:77-257.

Neck, R. W. 1983. Foodplant Ecology and Geographical Range of the Colorado Potato Beetle and a Related Species (*Leptinotarsa* ssp.) (Coleoptera: Chrysomelidae) Coleopt. Bull. 37(2): 177-182.

Oliver, A. G. 1791. Encyclopedia methodique. Histoire naturelle. Insectes. 5(2):691.

Opinion Number 1290, Bulletin of Zoological Nomenclature, Vol 42, Part 1, April 2, 1985. Conservation of *Leptinotarsa* Chevrolat 1837 (Insecta: Coleoptera), p. 21-23.

Parker, K. F. 1972. An Illustrated Guide to Arizona Weeds. The University of Arizona Press, Tucson, Arizona. 338 p.

Pope, R. D. and Madge, R. B. 1984. The "when" and "why" of the "Colorado potato beetle." Antennae, Bull. Ent. Soc. London 8(4):175-177.

Powell, E. F. 1941. Relationships within the family Chrysomelidae (Coleoptera) as indicated by the male genitalia of certain species. American Midl. Nat. 25:148-195.

Riley, C. V. 1863. Larvae of the ten-striped spearman. Prairie Farmer, n.s. 12:85-86.

Riley, C. V. 1867. The Colorado Potato Beetle. Prairie Farmer: 20:389.

Rogers, W. F. 1854. Synopsis of species of *Chrysomela* and allied genera inhabiting the United States. Proc. Acad. Nat. Sci. Philadelphia 8:29-39.

Say, T. 1824. Descriptions of coleopterous insects collected in the late expedition to the Rocky Mountains, performed by order of Mr. Calhoun, Secretary of War, under the command of Major Long. J. Acad. Nat. Sci. Philadelphia. 3(2):239-282, 298-331, 403-463.

Schaeffer, C. F. A. 1906. On new and known genera and species of the family Chrysomelidae. Sci. Bull. Mus. Brooklyn Inst. Arts Sci. 1(9):221-254.

Schroder, R. F. W. Athanas, M. M., Puttler, B. 1985. Propagation of the Colorado potato beetle, *Leptinotarsa decemlineata* (Coleoptera: Chrysomelidae) egg parasite *Edovum puttleri* (Hymenoptera: Eulophidae). Entomophaga 30(1):69-72.

Schroder, R. F. W. and Athanas, M. M. 1985. Review of Research on *Edovum puttleri* Grissell, egg parasite of the Colorado potato beetle. In Proceedings of the Symposium on the Colorado Potato Beetle,

XVIIth International Congress. D. Ferro and R. Voss (ed.), Mass. Ag. Exp. Sta. Res. Bull. No. 704.

Sorokin, N. S. 1982. Ground Beetles (Coleoptera: Carbidae) as Natural Enemies of the Colorado Potato Beetle, *Leptinotarsa decemlineata*. Entomol. Rev. 60(2), 1981. 44-52.

Stal, C. 1858. Till kannedomen om Amerikus chrysomeliner. Ofv. Svenska Vet.-Akad. Forh. 15:469-478.

Stal, C. 1859. Bidrag till kannedomen om Amerikas chrysomeliner. Ofv. Svenska Vet.-Akad. Forh. 16:305-326.

Stal, C. 1860. Till kannedomen on Chrysomelidae. Ofv. Svenska Vet.-Akad. Forh. 17:455-470.

Stal, C. 1862. Monographie des Chrysomelides de l'Amerique. I. Nov. Act. Soc. Sci. Upsaliensis. Ser. 3, 4:1-86.

Stal, C. 1863. Monographie des Chrysomelides de l'Amerique. II. Nov. Act. Soc. Sci. Upsaliensis. Ser. 3, 4:87-176.

Stal, C. 1865. Monographie des Chrysomelides de l'Amerique. III. Nov. Act. Soc. Sci. Upsaliensis. Ser. 3, 5:175-365.

Sturm, J. 1843. Catalog der Kaefer-Sammling von Jacob Sturm. Nurnberg. 386 pp.

Suffrian, E. 1858. (German translation and annotation of) F. W. Rogers, Ubersicht der in den Verein. Staaten von Nord-Amerika einheimischen Chrysomelen. Stettiner Ent. Zeitung. 19:237-278.

Tower, W. L. 1906. An investigation of Evolution in chrysomelid beetles of the genus *Leptinotarsa*. Carnegie Institution of Washington, publication no. 48, 320 pp.

Tower, W. L. 1918. The mechanism of evolution in *Leptinotarsa*. Carnegie Institution of Washington. Publication no. 263, 384 pp.

Townsend, C. H. 1903. Brownsville, Texas, Coleoptera. Trans. Texas Acad. Sci. 10:82-84.

Tuxen, S. L. 1970 (2nd rev. ed.) Taxonomist's Glossary of Genitalia in Insects. Scandinavian University Books. Munksgaard, Copenhagen. 359 pp.

Wagner, W. H. Jr. 1984. Applications of the Concepts of Groundplan Divergence, p. 95-118. In Cladistics: Perspectives on the Reconstruction of Evolutionary History. Thomas Duncan and Tod F. Stuessy (eds.) New York: Columbia University Press. 312 pp.

Walsh, B. D. 1863. The ten-striped spearman, Prairie Farmer n.s. 11:356.

Walsh, B. D. 1865. The New Potato-bug, and its Natural History. The Practical Entomologist 1:1-4.

Weise, J. 1913. Synonymische Bemerkungen. Wiener Ent. Zeitung. 32:17-18. Weise, J. 1916. Coleopterum Catalogus, Chrysomelidae: Chrysomelinae, pars 68, 225 pp. Junk.

Westhoff, F. 1878. Zur Speciesfrage des Kartoffelkafers. Ent. Nachrichten 4(9):113-118.

White, R. E. and Jacques, R. L. 1974. *Polygramma* Chevrolat, 1837: Proposed suppression under the plenary powers so as to conserve *Leptinotarsa* Stal, 1854 (Coleoptera, Chrysomelidae). Bull. Zool. Nomencl., Vol. 31, Pt. 3, 144-145.

Wilcox, J. A. 1954. Leaf beetles of Ohio (Coleoptera: Chrysomelidae). Ohio Biol. Surv. Bull., 43:353-506.

Wilcox, J. A. 1972. A Review of the North American Chrysomeline Leaf Beetles (Coleoptera: Chrysomelidae). Bulletin 421, State Education Department, New York State Museum, Albany, New York. 33 p.

Wilcox, J. A. 1979. Leaf beetle host plants in Northeastern North America. (coleoptera: Chrysomelidae). Gainesville, Florida: Flora and Fauna Publications, 30 p.

INDEX

ABOUT THE AUTHOR . . .

Richard L. Jacques, Ph.D. (Purdue University) was a Professor of Biology at Fairleigh Dickinson University in Rutherford, New Jersey and a Research Associate at the Florida State Collection of Arthropods in Gainesville, Florida. He taught both entomology and zoology at Fairleigh Dickinson University for the past fifteen years and was an assistant dean for the College of Science and Engineering. He was a member of the Entomological Society of America, Coleopterist's Society, New York Entomological Society and was a Registered Professional Entomologist. He was the co-author (with Ross H. Arnett, Jr.) of *Simon & Schuster's Guide to Insects* and *Insect Life: A Field Entomology Manual for the Amateur Naturalist.* He was born in New Jersey (1945) and died January 1, 1988.

Printed in the United States
by Baker & Taylor Publisher Services